Lecture Notes in Mathematics

Edited by A. Dold and B. Eckmann

467

Matts R. Essén

The cos $\pi\lambda$ Theorem

With a paper by Christer Borell

Springer-Verlag
Berlin · Heidelberg · New York 1975

Author
Prof. Matts R. Essén
Royal Institute of Technology
Division of Mathematics
Stockholm 70/Sweden

Library of Congress Cataloging in Publication Data

Essén, M R 1932-
The cos $\pi\lambda$ theorem.

(Lecture notes in mathematics ; v. 467)
Bibliography: p.
Includes index.
1. Potential, Theory of. 2. Functions, Subharmonic. 3. Inequalities (Mathematics) I. Title.
II. Series: Lecture notes in mathematics (Berlin) ;
v. 467.
QA3.L28 no.467 [QA404.7] 510'.8s [515'.7] 75-17547

AMS Subject Classifications (1970): 31-02, 31A05, 31A10, 31A15, 31B05, 34A40

ISBN 3-540-07176-8 Springer-Verlag Berlin · Heidelberg · New York
ISBN 0-387-07176-8 Springer-Verlag New York · Heidelberg · Berlin

Printed in Germany

Offsetdruck: Julius Beltz, Hemsbach/Bergstr.

The first eight sections of these notes are based on the second half of a course given at the University of Minnesota during the fall quarter of 1972 and at the University of Kentucky during the spring semester of 1973. During the first half of the course, an introduction to the theory of subharmonic functions was given, based on Chapters 4 and 5 in Heins (the name of an author refers to the corresponding book in the list of references). The purpose of the second half was to give a survey of recent results of Phragmén-Lindelöf type on the growth at infinity of functions subharmonic in $\underline{R}^2$. Many of these results can be extended to $\underline{R}^d$, $d \geq 3$, as is clear from the references.

The starting-point for the work reported here is the $\cos\pi\lambda$-theorem of Kjellberg [3]. For the history of the subject up to 1948, see Kjellberg [1]. The proofs given in these notes are sometimes different from those of the original papers. I have throughout used the approach of Hellsten-Kjellberg-Norstad [1] and of Essén ([2b], Lemma 3.1) (for further details, cf. sections 1 and 2). Apart from being able to use the same technique in the proofs of several known results, we are also able to extend certain $\cos\pi\lambda$-results to the situation considered by A. Baernstein [2]: the results thus obtained are new (cf. section 8).

Sections 1-4 contain results which have previously appeared elsewhere. In sections 3 and 4, I have avoided certain technical difficulties by only discussing $\underline{R}^2$. Sections 5-8 are new. Section 8 is joint work of the author and J. Lewis.

The essential tools are the Riesz representation formula for subharmonic functions and some properties of convolution inequalities. For those interested in the theory in $\underline{R}^d$, $d \geq 3$, I would like to mention that eigenfunction expansions of the Green or the Neumann function in a cone are also used.

I want to thank the Departments of Mathematics at the University of Minnesota and the University of Kentucky for the opportunity to give these lectures. I am also grateful to the chairman at the second university, Professor R. H. Cox for making it possible to get (a preliminary version of) these notes typed in Lexington.

Lexington, May 8, 1973

When I was working on the first eight sections of these notes during the spring of 1973, I received a preprint of the paper [2] of A. Baernstein. I found his extension of the $\cos\pi\lambda$-theorem very interesting and included part of his work in section 2: it also inspired J. Lewis and myself to the work reported in section 8. At the conference on Classical Function Theory in Canterbury in July 1973, A. Baernstein gave further applications of his rearrangement theorem for subharmonic functions. This time, he obtained new results on univalent functions and harmonic measures. At a series of seminars at the Mittag-Leffler Institute during the fall of 1973, I gave a survey of the work of Baernstein as well as an example of how the Baernstein technique could be extended to $\underline{R}^d$, $d \geq 3$. Most of this material is given in section 9.

At these seminars, it was noticed by Dr. Christer Borell that a result of Baernstein on estimates of harmonic measures could be extended to $\underline{R}^d$. This paper is also included in these lecture notes.

I am grateful to Albert Baernstein for letting me have preprints of his important papers in this area of research. My thanks are also due to Mrs. Beverly Mullins in Lexington and Mrs. Anna-Maria Johansson in Stockholm for their excellent typing. Finally, I would like to mention my wife Agneta without whose support and patience this work would not have been possible.

Stockholm, February 20, 1974

Matts Essén

In sections 1-3, proofs of results are given only in the case $0 < \lambda < 1/2$. The theorems are valid when $0 < \lambda < 1$. Similarly in section 4, we discuss only the case $\lambda_0 \geq 1/2$ in spite of the fact that there are results of this type also when $0 < \lambda_0 < 1/2$. Still, the proofs given here contain all the essential ideas which are needed in a complete discussion. It is not necessary for the understanding of these lecture notes to check that this statement is correct. However, it is easy for anyone who has read these notes and who wishes to know more to find the missing details in the original papers.

Let me also mention that so far, generalizations of the results in sections 1-3 to higher dimensions exist only in the case analogous to $0 < \lambda < 1/2$.

Notation. $\underline{R}$ is the real axis, $\underline{C}$ is the complex plane.

$$\Delta(a,r) = \{z \in \underline{C} : |z - a| < r\} \quad ,$$
$$C(a,r) = \{z \in \underline{C} : |z - a| = r\} \quad .$$

If u is subharmonic in a region $\Omega \subset \underline{C}$, let

$$M(r,u) = \sup u(z), \quad |z| = r, \quad z \in \Omega \quad ,$$
$$m(r,u) = \inf u(z), \quad |z| = r, \quad z \in \Omega \quad .$$

If it is clear from the context what subharmonic function we consider, we sometimes write $M(r)$ instead of $M(r,u)$. This remark applies also to other functionals than $M(r)$ and $m(r)$. Let $\partial\Omega$ be the boundary of Ω. If $\zeta \in \partial\Omega$ and u is subharmonic in Ω, define

$$u(\zeta) = \limsup u(z), \; z \to \zeta, \quad z \in \Omega \; .$$

If the region Ω is unbounded, we also introduce the order $\rho_0 = \limsup_{r \to \infty} \log M(r,u)/\log r$ and the lower order $\lambda_0 = \liminf_{r \to \infty} \log M(r,u)/\log r$ of u.

Contents

1. The Hellsten - Kjellberg - Norstad inequality.

Let u be subharmonic in $\Delta(0,R)$. Let $m(r,u) = m(r) = \inf u(z)$, $|z| = r$, $M(r,u) = M(r) = \sup u(z)$, $|z| = r$ and $M(R) = \sup M(r)$, $0 \le r < R$. Assume that λ is given, $0 < \lambda < 1$. If

$$(1.1) \qquad m(r) \le \cos\pi\lambda \; M(r), \; 0 < r < R \quad ,$$

what can we say about the growth of M(r) in the interval (0,R) ? If $\lambda = 1/2$ in (1.1), it is known that

$$M(r) \le \frac{4M(R)}{\pi} \arctan \sqrt{r/R} \quad .$$

This is the Milloux-Schmidt inequality (see Heins, p. 108). In the general case, the following result is true.

Theorem 1.1: If $0 < M(R) < \infty$, $0 < \lambda < 1$ and (1.1) is true, there exists an extremal subharmonic function in $\Delta(0,R)$

$$U(z) = \mathrm{Re}\{\frac{2M(R)}{\pi} \tan \frac{\pi\lambda}{2} \int_0^{z/R} (t^{\lambda-1} - t^{1-\lambda}) \, (1-t^2)^{-1} dt\}, \; |\arg z| \le \pi \; ,$$

for which there is equality in (1.1) and for which

$$M(r,u) \le M(r,U) = U(r) \quad .$$

Corollary: $r^{-\lambda} M(r) \le C_0(\lambda) \, R^{-\lambda} M(R)$, $0 < r < R$, where

$C_0(\lambda) = (\tan \frac{\pi\lambda}{2})(\frac{\pi\lambda}{2})^{-1}$. The constant $C_0(\lambda)$ is best possible.

Remarks:

a. If $M(R) \le 0$, (1.1) is trivially satisfied.

b. It follows from (1.1) that $u(0) \le 0$. If we assume that u is continuous in some disk centered at the origin or if $u(z) \to -\infty$, $z \to 0$, this is readily seen. A proof that $u(0) \le 0$ which avoids these restrictions is given at the end of section 1. It is accessible to those who know potential theory as presented e.g. in Helms. Other readers are recommended to go on assuming that u is continuous near the origin.

In the sequel, we shall for simplicity only discuss the case $0 < \lambda \leq 1/2$. A complete discussion of the case $0 < \lambda < 1$ is in the paper of Hellsten, Kjellberg and Norstad. A subharmonic function which is bounded above in a disk $\Delta(0,R)$ can be written in the form

$$u(z) = u_1(z) + u_2(z) .$$

There exist nonnegative measures ν and μ such that

$$u_1(z) = \iint_{|\zeta| < R} \log \left|\frac{R(z-\zeta)}{R^2 - z\bar{\zeta}}\right| d\mu(\zeta) ,$$

$$u_2(z) = M(R) - \frac{1}{2\pi} \int_{-\pi}^{\pi} \frac{R^2 - |z|^2}{|Re^{i\theta} - z|^2} d\nu(Re^{i\theta}) ,$$

(cf. Tsuji IV.10). We now project the Riesz mass onto the negative real axis, i.e., we consider an associated subharmonic function

$$u^*(z) = u_1^*(z) + u_2^*(z) ,$$

where

$$u_1^*(z) = \iint_{|\zeta| < R} \log \left|\frac{R(z+|\zeta|)}{R^2 + z|\zeta|}\right| d\mu(\zeta) ,$$

$$u_2^*(z) = M(R) - \frac{1}{2\pi} \frac{R^2 - |z|^2}{|R+z|^2} \int_{-\pi}^{\pi} d\nu(Re^{i\theta}) .$$

The function u^* has its Riesz mass concentrated to the segment $[-R,0]$ and is harmonic in $\Delta(0,R)\setminus[-R,0]$. Furthermore,

$$u^*(-r) = m(r,u^*) \leq m(r,u) \leq M(r,u \leq M(r,u^*) = u^*(r) \tag{1.2}$$

This is clear since

$$\frac{R|r - |\zeta||}{R^2 - r|\zeta|} \leq \left|\frac{R(z - \zeta)}{R^2 - z\bar{\zeta}}\right| \leq \frac{R(r + |\zeta|)}{R^2 + r|\zeta|} .$$

Using (1.1) and (1.2), we see that

$$u^*(-r) \le \cos\pi\lambda\ u^*(r) . \tag{1.3}$$

Now let H be a harmonic function bounded from above in $\Delta(0,R)\setminus[-R,0]$ which is representable as an integral of its boundary values, and is such that $H(z) = H(\bar{z})$.
Then

$$H(r) = \int_0^R Q(r,t)H(-t)\,dt + \int_{-\pi}^{\pi} T(r,\varphi)H(Re^{i\varphi})d\varphi = I_1 + I_2 , \tag{1.4}$$

where

$$Q(r,t) = \frac{1}{\pi}\sqrt{r/t}\ \{(t+r)^{-1} - R(R^2 + rt)^{-1}\} ,$$

$$T(r,\varphi) = \frac{1}{\pi}\sqrt{Rr}\ (R - r)\cos(\varphi/2)\ \{R^2 + r^2 - 2Rr\cos\varphi\}^{-1} ,$$

(cf. Boas, p. 2). Replacing H by u^* in (1.4) and using (1.2), we see that

$$I_1 \le \cos\pi\lambda \int_0^R Q(r,t)M(t,u)\,dt \le \cos\pi\lambda \int_0^R Q(r,t)\,u^*(t)\,dt ,$$

$$I_2 \le M(R) \int_{-\pi}^{\pi} T(r,\varphi)\,d\varphi = \frac{4M(R)}{\pi}\arctan\sqrt{r/R} ,$$

and hence that

$$u^*(r) \le \cos\pi\lambda \int_0^R Q(r,t)\,u^*(t)\,dt + \frac{4}{\pi}M(R)\arctan\sqrt{r/R} , \tag{1.5a}$$

$$u^*(r) \le \cos\pi\lambda \int_0^R Q(r,t)\,M(t,u)\,dt + \frac{4}{\pi}M(R)\arctan\sqrt{r/R} . \tag{1.5b}$$

In the present discussion, we are interested in (1.5a). Formula (1.5b) is included for further reference.

We require the following lemma. Let C[0,R] be the space of continuous functions on [0,R] under the maximum norm.

Lemma: Let $T = C[0,R] \to C[0,R]$ be an integral operator with a non-negative kernel such that $\| T \| < 1$.

a) If $h \in C[0,R]$ is given, there exists a unique solution in $C[0,R]$ of the equation

$$U = TU + h .$$

b) If U, W and h are nonnegative functions in $C[0,R]$ such that

$$U = TU + h , \quad W \leq TW + h ,$$

then $W \leq U$.

Proof: a) It is easily seen that $U = \sum_0^\infty T^n h$ converges in $C[0,R]$ and is the unique solution.

b) It follows from the assumption that $(U - W) = T(U - W) + g$ where $g \in C[0,R]$ and is nonnegative. Hence the unique solution of this equation is nonnegative, and b) is proved.

We claim that the operator T defined by

$$TU(r) = \cos\pi\lambda \int_0^R Q(r,t)\, U(t)\, dt$$

has the properties demanded in the lemma if $0 < \lambda \leq 1/2$. It is easy to see that if $U \in C[0,R]$, $TU \in C[0,R]$. The only property of T which is not obvious is that $\| T \| < 1$. To prove this, we replace H by the constant 1 in (1.4) and obtain $\int_0^R Q(r,t)\, dt < 1$, $0 < r < R$. Thus, if $\| U \|_\infty \leq 1$, $\| TU \|_\infty \leq \cos\pi\lambda < 1$, and the proof is complete.

We now use the lemma to compare u^* which satisfies (1.5) to the unique solution U of

$$U(r) = TU(r) + \frac{4}{\pi} M(R) \arctan \sqrt{r/R} . \tag{1.6}$$

There is one difficulty: u^* restricted to $[0,R]$ may not be an element in $C[0,R]$ since it can occur that $u^*(0) = -\infty$. If this is the case, let us introduce $v(r) = \max(u^*(r),0)$. It is clear that (1.5a) is valid with u^* replaced by v and that $v \in C[0,R]$. By the second part of the lemma, $v \leq U$ and hence $M(r,u) \leq u^*(r) \leq v(r) \leq U(r)$ (cf. (1.2)).

Thus we have an estimate of $M(r,u)$. It remains to get more information on U. To do this, we shall now, following Norstad, solve equation (1.6). Putting $r = Re^{-x}$, $t = Re^{-s}$ in (1.6), we obtain

$$(1.7) \qquad \Phi(x) = \int_0^\infty \{K(x-s) - K(x+s)\}\,\Phi(s)ds + g(x),\ x > 0,$$

where

$$\Phi(x) = U(Re^{-x}) .$$

$$K(x) = \frac{\cos\pi\lambda}{\pi}\,(2\cosh(x/2))^{-1}$$

and

$$g(x) = \frac{4M(R)}{\pi}\arctan e^{-x/2},\ x > 0 .$$

Extending Φ and g to the whole real line so that the new functions are odd, it follows from (1.7) that

$$\Phi(x) = \int_{-\infty}^{\infty} K(x-s)\,\Phi(s)ds + g(x),\ x \in \underline{R} .$$

(The extended functions are also denoted by Φ and g.)

Introducing Fourier transforms,

$$\hat{\Phi}(t) = \int_{-\infty}^{\infty} \Phi(x)\,e^{itx}dx,\ \hat{g}(t) = \int_{-\infty}^{\infty} g(x)\,e^{itx}dx,$$

we proceed formally: $\hat{\Phi}(t) = \hat{K}(t)\,\hat{\Phi}(t) + \hat{g}(t)$,

and thus

$$\Phi(x) = \frac{1}{2\pi} \int_{-\infty}^{\infty} \hat{g}(t)(1 - \hat{K}(t))^{-1} e^{-ixt} dt .$$

This integral is absolutely convergent and can be evaluated using the calculus of residues. We obtain the formula for U given in Theorem 1.1 which gives us the unique solution of equation (1.6) in C[0,R].

Details in this evaluation of $\Phi(x) = U(Re^{-x})$ are given at the end of section 1.

We have now proved most of Theorem 1.1. It remains to study $U = \text{Re } w$, where

$$w(z) = \frac{2M(R)}{\pi} \tan \frac{\pi\lambda}{2} \int_0^{z/R} (t^{\lambda-1} - t^{1-\lambda})(1 - t^2)^{-1} dt ,$$

which is analytic in $\Delta(0,R) \setminus [-R,0]$. We have the following properties of U:

a) $U(-r) = \cos\pi\lambda \; U(r) \; , \; 0 \le r < R$.

b) The variation of $w(z)$ on $\{z = Re^{i\theta} \; , \; |\theta| < \pi\}$, is purely imaginary, i.e. $\text{Re } w(z)$ is constant on this arc. Thus

$$U(Re^{i\theta}) = U(R) = M(R) \; , \; |\theta| < \pi .$$

c) $U(z) = U(\bar{z})$.

d) We claim that U is subharmonic in $\Delta(0,R)$. Since U is harmonic in $\Delta(0,R) \setminus [-R,0]$, it suffices to study points on the negative real axis. A calculation shows that at each point on the segment, the inner normal derivatives in both directions are positive (and of course equal due to c). Continuation of $U(z)$ from above the segment gives, in a disc $|z + r| < \delta$, a harmonic function which is less than $U(z)$ in the lower half of the disc. Thus a local condition for subharmonicity is satisfied at $z = -r$. Since $U(0) = 0$, a local condition for subharmonicity is satisfied also at the origin. The proof is complete.

The corollary is a consequence of the inequality

$$(t^{\lambda-1} - t^{1-\lambda})(1 - t^2)^{-1} \le t^{\lambda-1} \; , \; 0 < t < 1 .$$

It is clear that the constant $C_o(\lambda)$ is best possible.

In the sequel, we shall a number of times meet a problem of the following type: let v be a nonnegative, upper semicontinuous function which is 0 near the origin and is such that

$$(1.5)\quad v(r) \leq \cos\pi\lambda \int_0^R Q(r,t)\, v(t)dt + \text{Const. } M(R)\sqrt{r/R}\ ,\quad 0 < r \leq R\ .$$

Does there exist a constant such that

$$(1.8)\qquad r^{-\lambda}v(r) \leq \text{Const. } R^{-\lambda}M(R)\ ,\quad 0 < r < R\ .$$

Theorem 1.1 gives a result of this type with the best possible constant. Without solving the associated integral equation, we can deduce (1.8) in the following simple way (which does not give the best constant) (cf. Essén [2b], Lemma 3.1). We shall use this technique later for other kernels than Q.

It follows from (1.5) that

$$(1.9)\qquad r^{-\lambda}v(r) \leq \frac{\cos\pi\lambda}{\pi} \int_0^R t^{-\lambda}v(t)(r/t)^{1/2-\lambda}(1 + r/t)^{-1}\,\frac{dt}{t} +$$

$$+ \text{Const. } R^{-\lambda}M(R)\ (r/R)^{1/2-\lambda}\ .$$

Let $M = \sup_{0 \leq r \leq R} r^{-\lambda}v(r) = r_o^{-\lambda}v(r_o)$.

The maximum is assumed since v is upper semicontinuous and zero near the origin. Putting $r = r_o$ in (1.9), we see that

$$M \leq M\,\frac{\cos\pi\lambda}{\pi} \int_0^{R/r_o} s^{\lambda-1/2}(1 + s)^{-1}ds + \text{Const. } R^{-\lambda}M(R)(r_o/R)^{1/2-\lambda}\ ,$$

$$M\,\frac{\cos\pi\lambda}{\pi} \int_{R/r_o}^{\infty} s^{\lambda-1/2}(1 + s)^{-1}ds \leq \text{Const. } R^{-\lambda}M(R)(r_o/R)^{1/2-\lambda}\ ,$$

and thus that

$$(1.10)\qquad M \leq \text{Const. } 2^{1/2-\lambda}\,\frac{\pi(1/2 - \lambda)}{\cos\pi\lambda}\,R^{-\lambda}\,M(R)\ .$$

We have here used that

$$\int_0^\infty t^{\lambda-1/2}(1+t)^{-1}dt = \frac{\pi}{\cos\pi\lambda} \quad , \quad \int_\rho^\infty t^{\lambda-1/2}(1+t)^{-1}dt \geq \frac{(1+\rho)^{\lambda-1/2}}{1/2-\lambda} .$$

The result we want is (1.8) which is immediate from (1.10) and the definition of M.

Remark 1: We have claimed that it follows from (1.1) that $u(0) \leq 0$. This is proved in the following way.

By definition, the fine topology on $\underline{R}^2$ is the smallest topology making all subharmonic functions continuous in the extended sense (cf. Helms p. 207).

A set E is thin at z if z is not a fine limit point of E (cf. Helms p. 209).

Let $\varepsilon > 0$ be given. Then $Q_\varepsilon = \{z: u(z) > u(0) - \varepsilon\}$ is a fine neighbourhood of the origin. If $u(0) = -\infty$, there is nothing to prove. We can assume that u(0) is finite. Since

$$\lim\sup u(z) < u(0) \ , \ z \to 0 \ , \ z \notin Q_\varepsilon \ ,$$

it follows from Theorem 10.3 in Helms that the complement of Q_ε is thin at the origin. Thus there exists a sequence of circles $\{C(0,r_\nu)\}$, $r_\nu \to 0$, $\nu \to \infty$, such that $C(0,r_\nu) \subset Q_\varepsilon$ for all ν (Theorem 10.14 in Helms). Thus when $\nu \to \infty$, we see that

$$u(0) - \varepsilon \leq m(r_\nu) \leq \cos \pi\lambda M(r_\nu) \to \cos \pi\lambda u(0) \ ,$$

or equivalently, $u(0) \leq \varepsilon\,(1 - \cos \pi\lambda)^{-1}$. Since $\varepsilon > 0$ is arbitrary, we have proved that $u(0) \leq 0$.

Remark 2: It remains to fill in the missing details in the solution of (1.6). We first note that

$$\hat{K}(t) = \cos \pi\lambda \ / \ \cosh \pi t \ ,$$

$$\hat{g}(t) = 2\,i\,M(R)(1 - \frac{1}{\cosh \pi t})t^{-1} \ .$$

The integral which we wish to evaluate is

$$\phi(x) = (i/\pi) \int_{-\infty}^{\infty} M(R) \frac{\cosh \pi t - 1}{\cosh \pi t - \cos \pi\lambda} e^{-ixt} dt/t .$$

We evaluate $\phi(x)$ when $x > 0$. Integrating over an interval on the real axis and a half-circle in the lower half-plane, we find that the integral has simple poles at $\{(\lambda - 2n)i\}_1^\infty$ and $\{-(2n + \lambda)i\}_0^\infty$. Applying the residue theorem, we obtain

$$\phi(x) = 2\, M(R) \frac{1 - \cos \pi\lambda}{\pi \sin \pi\lambda} \left| \sum_{n=0}^{\infty} \frac{e^{-x(\lambda + 2n)}}{\lambda + 2n} - \sum_{1}^{\infty} \frac{e^{-x(2n - \lambda)}}{2n - \lambda} \right| ,$$

$$U(r) = 2\, M(R) \frac{1 - \cos \pi\lambda}{\pi \sin \pi\lambda} \left| \sum_{n=0}^{\infty} \frac{(r/R)^{2n + \lambda}}{2n + \lambda} - \sum_{n=1}^{\infty} \frac{(r/R)^{2n - \lambda}}{2n - \lambda} \right| ,$$

which is equivalent to

$$U(r) = \frac{2\, M(R)}{\pi} \tan \frac{\pi\lambda}{2} \int_0^{r/R} \frac{t^{\lambda - 1} - t^{1 - \lambda}}{1 - t^2}\, dt .$$

2. The $\cos\pi\lambda$-theorem

Theorem 2.1: Let u be a non-constant subharmonic function in C and let λ be given, $0 < \lambda < 1$. If

$$(2.1) \quad m(r) \leq \cos\pi\lambda\, M(r), \quad r \geq r_0,$$

then $\lim_{r\to\infty} r^{-\lambda} M(r) = \alpha$ exists, $0 < \alpha \leq \infty$.

This result is due to Kjellberg [3]. The case $\lambda = 1/2$ is in Heins [1]. The proof we give is not the original one given by Kjellberg. Instead, we shall use the Hellsten-Kjellberg-Norstad inequality discussed in § 1. Kjellberg's original technique has been successfully applied to other problems in Edrei [1] and Baernstein [2]. We shall discuss the remarkable result of Baernstein later in this section.

For simplicity, we give the proof of Theorem 2.1 in the case $0 < \lambda \leq 1/2$ only.

a) Let $u^+ = \max(u,0)$. We can assume that (2.1) is true for all $r > 0$, that u is nonnegative and that u is 0 in a neighborhood of the origin. If this is not true, we consider $u_1 = (u - M(r_0,u))^+$ in place of u. It is easily checked that u_1 fulfills all the conditions mentioned above.

b) We apply the corollary of Theorem 1.1 in the disc $\Delta(0,R)$. It follows that

$$r^{-\lambda}M(r) \leq C_0(\lambda)\, R^{-\lambda}M(R), \quad 0 < r \leq R.$$

Let $\beta = \liminf_{r\to\infty} R^{-\lambda}(M(R)$. It follows that $\limsup_{r\to\infty} r^{-\lambda}M(r) \leq C_0(\lambda)\beta$.

Thus either $\beta = \alpha = \infty$ or

$$(2.2) \quad M(R) = O(R^{\lambda}), \quad R \to \infty.$$

c) From now on, we assume that (2.2) holds. Writing $M(r,u) = M(\cdot,,$ it is clear from (2.1), (1.2) and (1.5b) that

$$(2.3) \qquad M(r) \leq \cos\pi\lambda \int_0^R Q(r,t)\, M(t)\, dt + \frac{4}{\pi} M(R) \arctan\sqrt{r/R} \, .$$

Letting $R \to \infty$ and dividing by r^λ, we see that

$$(2.4) \qquad r^{-\lambda}M(r) \leq \frac{\cos\pi\lambda}{\pi} \int_0^\infty (r/t)^{1/2-\lambda}(t + r)^{-1}\, t^{-\lambda}M(t)\, dt \, .$$

By the change of variables $r = e^x$, $t = e^y$, we obtain

$$(2.5) \qquad \Phi(x) \leq \Phi * K(x) = \int_{-\infty}^{\infty} \Phi(y)\, K(x - y)\, dy \, ,$$

where $\Phi(x) = e^{-\lambda x} M(e^x)$ and

$$(2.6) \qquad K(x) = (\cos\pi\lambda/\pi)\, e^{-\lambda x} \left(2\cosh \frac{x}{2}\right)^{-1} .$$

It follows from (2.5) that $\alpha = \lim_{x\to\infty} \Phi(x)$ exists and is positive, and this is the conclusion of Theorem 2.1. We shall several times meet inequalities similar to (2.5). Therefore, we state a general result from which the existence of the limit follows (cf. Essén [1]). We first need a definition. A function $f\colon \underline{R} \to \underline{R}$ is slowly decreasing if

$$\liminf_{h\to 0+} \{\liminf_{x\to\infty} (f(x+h) - f(x))\} \geq 0$$

(cf. Widder (Def. 9b, Ch. V)).

<u>Theorem 2.2</u>: <u>Let</u> $\Phi \in L^\infty$ <u>be slowly decreasing as</u> $x \to \infty$. <u>Let K be a non-negative function in</u> $L(-\infty , \infty)$ <u>such that</u>

$$(2.7) \qquad \int_{-\infty}^{\infty} K(x)dx = 1 \, , \quad \int_{-\infty}^{\infty} |x|\, K(x)\, dx < \infty \, ,$$

$$(2.8) \qquad m = \int_{-\infty}^{\infty} x\, K(x)\, dx \neq 0 .$$

If (2.5) holds, then $\lim_{x\to\infty} \Phi(x)$ *exists. If* $\lim_{x\to-\infty} \Phi(x) = 0$ *and if* $m < 0$, $\lim_{x\to\infty} \Phi(x) \geq 0$. *Equality occurs if and only if there is equality in (2.5). We also note that* $\int_{-\infty}^{\infty} (\Phi * K - \Phi)(x)\, dx < \infty$.

Remark: If there exists a function $\Phi \in L^\infty$ such that (2.5) holds and $\Phi \neq \Phi * K$, it follows that $m \neq 0$. However, instead of proving this fact, we shall check (2.8) directly when we wish to apply Theorem 2.2.

Before proving this theorem, let us use it to finish the proof of Theorem 2.1. If K is defined by (2.6), it is easy to check that (2.7 is true. Differentiating with respect to λ, we see that

$$m = \int_{-\infty}^{\infty} x\, K(x) = -\pi \tan \pi\lambda < 0 .$$

Since M(r) is nonnegative and (2.2) holds, we know that $\Phi(x) = e^{-\lambda x} M(e^x) \in L^\infty$ and that $\lim_{x\to-\infty} \Phi(x) = 0$. Let $h > 0$ be given. Since $M(e^x)$ is increasing,

$$\Phi(x+h) - \Phi(x) = e^{-\lambda(x+h)} M(e^{x+h}) - e^{-\lambda x} M(e^x) \geq \Phi(x)(e^{-\lambda h} - 1) ,$$

and it follows that $\underline{\lim}_{h\to 0}\, \underline{\lim}_{x\to\infty} (\Phi(x+h) - \Phi(x)) \geq 0$, e.g., Φ is slowly decreasing. Hence the functions Φ and K which appeared in the proof of Theorem 2.1 satisfy the assumptions of Theorem 2.2. The existence of the positive limit $\alpha = \lim_{x\to\infty} \Phi(x) = \lim_{r\to\infty} r^{-\lambda} M(r)$ is thus a consequence of Theorem 2.2.

Proof of Theorem 2.2: We set $\Phi - \Phi * K = h \leq 0$. Integrating, we see that

$$(2.9) \qquad \Phi * N(x) - \Phi * N(0) = \int_0^x h(y)\, dy ,$$

where

$$N(x) = \begin{cases} \int_x^\infty K(y)\,dy , & x > 0 , \\ -\int_{-\infty}^x K(y)\,dy , & x < 0 . \end{cases}$$

By (2.7), $N \in L(-\infty, \infty)$ and we have assumed that $\Phi \in L^\infty$. Consequently,

$$\left|\int_0^x h(y)\,dy\right| \leq \text{Const.} \int_{-\infty}^{\infty} |N(y)|\,dy < \infty .$$

Since h is non-positive, we conclude that $h \in L(-\infty, \infty)$. It follows from (2.9) that $\lim_{x\to\infty} \Phi * N(x)$ exists. We now need the following result (Pitt's Tauberian theorem, cf. Theorem 10a, ch. V in Widder):

Let $f \in L(\underline{R})$ be such that $\hat{f}(t) \neq 0$, $t \in \underline{R}$. Let $\varphi \in L^\infty(\underline{R})$ be slowly decreasing. If $\lim_{x\to\infty} \varphi * f(x) = a$ exists, then $\lim_{x\to\infty} \varphi(x) = a/\hat{f}(0)$.

Since Φ is slowly decreasing, the existence of $\lim_{x\to\infty} \Phi(x)$ is proved if $\hat{N}(t) \neq 0$, $t \in R$. But $\hat{N}(t) = (1 - \hat{K}(t))/(it)$, $t \neq 0$, and $\hat{N}(0) = m \neq 0$. Since $|\hat{K}(t)| < 1$, $t \neq 0$, we have proved the existence of $\lim_{x\to\infty} \Phi(x)$. It is easy to show that if

$$\lim_{x\to-\infty} \Phi(x) = 0 , \text{ then } \lim_{x\to\infty} \Phi * N(x) = \int_{-\infty}^{\infty} h(y)\,dy = \hat{h}(0) \leq 0 .$$

Once more applying Pitt's theorem, it follows that $\lim_{x\to\infty} \Phi(x) = \hat{h}(0)/m$, and Theorem 2.2 is proved.

A remarkable generalization of the $\cos\pi\lambda$ theorem has recently been given by A. Baernstein [2]:

Theorem 2.3: Let u be a non-constant subharmonic function in $\mathbb{C}$. Let β and λ be numbers such that $0 < \lambda < \infty$, $0 < \beta \leq \pi$, $\beta\lambda < \pi$.

If for any interval J of length 2β and for all $r > r_0$,

(2.10) $$\inf_{\theta \in J} u\,(re^{i\theta}) \leq \cos\beta\lambda\; M(r) ,$$

then $\lim_{r\to\infty} r^{-\lambda} M(r) = \alpha$ exists, $0 < \alpha \leq \infty$.

At this stage, we cannot give all details in the rather long proof of this result. Baernstein's theory will be discussed in section 9. We would like to mention how the techniques emphasized in these lecture notes can replace certain parts of Baernstein's argument (which in their turn are based on Kjellberg's original paper [3]). In this way we shall later be able to extend several $\cos\pi\lambda$-results to the situation considered by Baernstein.

Baernstein introduces the following concepts:

let $$u(r,\theta,\varphi) = \int_{-\theta}^{\theta} u(re^{i(w\,+\,\varphi)})\, dw ,$$

(2.11) $$v(re^{i\theta}) = \sup_{\varphi} u(r,\theta,\varphi) .$$

For each fixed $re^{i\theta}$, there exists an interval $I(r)$ for which the sup in (2.11) is attained. Let $\nu(r) = \inf u(re^{i\omega})$, $\omega \in I(r)$. Baernstein proves that

a) $v(z)$ is subharmonic in $\{\mathrm{Im}z > 0\}$ and continuous on $\{\mathrm{Im}z \geq 0\}$, except perhaps at $z = 0$.

b) For each fixed $\beta \in (0,\pi)$, $v(re^{i\beta})$ is a non-decreasing convex function of $\log r$, $0 < r < \infty$.

Now, let $D_0 = \{z\colon 0 < |z| < R , 0 < \arg z < \beta\}$. Let H be the bounded harmonic function in D which has the following boundary values:

$$H(r) = 0 , \quad H(re^{i\beta}) = v(re^{i\beta}) , \quad 0 \leq r < R .$$

$$H(Re^{i\theta}) = \begin{cases} 2\pi M(R) \ , \ 0 < \theta < \beta/2 \ , \\ \\ 4\pi M(R) \ , \ \beta/2 < \theta < \beta \ . \end{cases}$$

Using the reflection principle, we extend the harmonic function H to $D_1 = \{z: 0 < |z| < R \ , \ |\arg z| < \beta\}$. It is clear that $H(\bar{z}) = -H(z) \ , \ z \in D_1$. Now consider the harmonic function $H_\theta(re^{i\theta}) = \frac{\partial}{\partial\theta} H(re^{i\theta})$. Baernstein proves the following two inequalities which are the key to his proof:

(2.12) $$H_\theta(r) \geq 2M(r) \ ,$$

(2.13) $$H_\theta(re^{i\beta}) + H_\theta(r) \leq 2\{\nu(r) + M(r)\} \ ,$$

(cf. (31), (32) in Baernstein [2]).

At this point, Baernstein uses the original argument of Kjellberg [3]. We prefer to use Theorem 1.1. First, we require the following estimate which depends on the fact that $v(re^{i\beta})$ is an increasing convex function of $\log r$ (a proof is given on pp. 70-73):

(2.14) $$H_\theta(\tfrac{1}{2} Re^{i\theta}) \leq \text{Const.}\ M(R) \ , \ |\theta| \leq \beta \ .$$

If we now map $D_1 \cap \Delta(0,R/2)$ onto a disk cut along the negative real axis, we find ourselves in a situation similar to the one discussed in § 1. Let $\gamma = \beta/\pi$, $R^* = (R/2)^{1/\gamma}$ and define g in $D_2 = \Delta(0,R^*) - [-R^*,0]$ by $g(z) = H_\theta(z^\gamma)$. By (2.12) - (2.14) ,

(2.12') $$g(\rho) \geq 2M(\rho^\gamma) \ , \ 0 < \rho \leq R^* \ ,$$

(2.13') $$g(-\rho) + g(\rho) \leq 2\{\nu(\rho^\gamma) + M(\rho^\gamma)\} \ , \ 0 < \rho \leq R^* \ ,$$

(2.14') $$g(R^* e^{i\theta}) \leq \text{Const.}\ M(R) \ , \ |\theta| \leq \pi \ .$$

By (2.10) and (2.13')

$$(2.15) \qquad g(-\rho) + g(\rho) \le 2(1 + \cos\beta\lambda)\, M(\rho^{\gamma}) \;, \; 0 < \rho \le R^* .$$

We also observe by (2.12') and (2.15),

$$(2.16) \qquad g(-\rho) + g(\rho) \le (1 + \cos\beta\lambda)\, g(\rho) \;, \; 0 < \rho \le R^* ,$$

$$(2.16') \qquad g(-\rho) \le \cos\beta\lambda\; g(\rho) .$$

In the statement of Theorem 1.1, let us replace λ by $\lambda' = \beta\lambda/\pi = \lambda\gamma$. If a function g is harmonic in $\Delta(0,R^*)\setminus[-R^*,0]$ and satisfies (2.14') and (2.16'), it is majorized on $[0,R]$ by an extremal Hellsten-Kjellberg-Norstad function, possibly multiplied by a constant only depending on the constant in (2.14'). Hence

$$\rho^{-\gamma\lambda}\, g(\rho) \le \text{Const. } (R^*)^{-\gamma\lambda}\, g(R^*) = \text{Const. } R^{-\lambda}\, M(R) \;, \; 0 < \rho \le R^* .$$

By putting $\rho^{\gamma} = r$ we see that

$$r^{-\lambda}\, M(r) \le \text{Const. } R^{-\lambda}\, M(R) \;, \; 0 < r \le R/2 .$$

Since $M(r)$ is increasing, this inequality is trivially true for $R/2 < r \le R$. Just as in the discussion in b) at the beginning of the proof of Theorem 2.1, it follows that either $\beta = \liminf\limits_{R\to\infty} R^{-\lambda} M(R) = \infty$ or $M(R) = O(R^{\lambda})$, $R \to \infty$. In the later case and if $0 < \gamma\lambda \le 1/2$, we repeat the discussion leading up to (2.3) and (2.4) to obtain

$$(2.4') \quad \rho^{-\gamma\lambda}\, M(\rho^{\gamma}) \le \frac{\cos\beta\lambda}{\pi} \int_0^{\infty} (\rho/t)^{1/2-\gamma\lambda}\, (t + \rho)^{-1}\, t^{-\gamma\lambda}\, M(t^{\lambda})\, dt .$$

The existence of $\lim\limits_{r\to\infty} r^{-\lambda} M(r) = \lim\limits_{\rho\to\infty} \rho^{-\gamma\lambda} M(\rho^{\gamma})$ is now immediate from Theorem 2.2.

<u>Remark</u>: If $1/2 < \gamma\lambda < 1$, (2.16') is in the previous discussion replaced by (2.16). One more trick is used in this case: we have to use the integral inequality (18) in Hellsten-Kjellberg-Norstad [1] or the integral inequality (3.5) in Essén [2b] rather than the simple inequality (17) in Hellsten-Kjellberg-Norstad [1] which is the one used above.

3. A generalization of the Ahlfors-Heins theorem

Theorem 3.1: Let u be subharmonic in $\{\mathrm{Re}\, z > 0\}$. Let λ be given, $0 < \lambda < 1/2$. If

$$u(iy) \leq \cos\pi\lambda \; M^+(|y|) \; , \; y \in \underline{R} \; , \; y \neq 0 \; , \tag{3.1}$$

$$u(0) < \infty \; , \tag{3.2}$$

(3.3) there exists z_0 , $\mathrm{Re}\, z_0 > 0$, such that $u(z_0) > 0$, then either $r^{-2\lambda} M(r) \to \infty$ or

$$\lim u(re^{i\theta})/(r^{2\lambda} \cos 2\lambda\theta) = \alpha \; , \; z \to \infty \; , \; z \notin E \; , \tag{3.4}$$

where $\alpha > 0$ and the exceptional set E can be covered by disks $\{B_i\}_1^\infty$ such that if r_i is the radius of B_i and R_i is the distance from the centre of B_i to the origin,

$$\sum_1^\infty (r_i/R_i) < \infty \; . \tag{3.5}$$

In the plane, this theorem is in fact true when $0 < \lambda < 1$ (cf. Essén [2], Lewis [3] and Essén-Lewis [1]). After the proof of Lemma 3.1, we only discuss the case $0 < \lambda < 1/2$. The case $\lambda = 1/2$ is in Ahlfors-Heins [1] except that their characterization of the exceptional set E is different. For the further history of the set E, cf. Hayman [1] and Azarin [1]. The method of covering the set E used in Theorem 3.1 is due to Azarin.

Theorem 3.1 also holds in circular cones in $\underline{R}^d$, $d \geq 3$ (cf. Essén-Lewis [1]). We conjecture that this result is true also in more general cones. The case when the subharmonic function has non-positive boundary values has been treated by J. Lelong-Ferrand [1] (in the plane, this is the theorem of Ahlfors-Heins mentioned above). Using her ideas, more can be said about the exceptional set E (cf. Essén-Jackson [1]).

We note that it follows from (3.1) and (3.3) that $M(r) \geq M(|z_0|) > 0, r > |z_0|$.

For convenience, let us here state two theorems which will be used in the sequel.

The Phragmén-Lindelöf maximum principle. Let u be subharmonic in a region Ω and let E be a countable proper subset of $\partial\Omega$. If $\sup u(z) < \infty$, $z \in \Omega$, and $u(\zeta) \leq M$, $\zeta \in \partial\Omega \setminus E$, then $\sup u(z) \leq M$, $z \in \Omega$. (cf. Heins p. 76).

The Phragmén-Lindelöf theorem.

Let u be subharmonic in $D = \{\mathrm{Re}\, z > 0\}$ and assume that $u(iy) \leq 0$, $y \in \underline{R}$. Then either $u \leq 0$ in D or else $\liminf M(r,u)/r > 0$, $r \to \infty$. (cf. Heins p. 111).

We need the following lemma (cf. Lewis [2]).

Lemma 3.1: Let Ω be a region contained in $\Delta(0,R)$ and assume that $\partial\Omega \cap C(0,r) \neq \emptyset$, $0 < r < R$. If λ is given, $0 < \lambda < 1$, if u is subharmonic in Ω , if $u(0) < \infty$ and if

(3.6) $u(\zeta) \leq \cos\pi\lambda \, M^+(|\zeta|)$, $\zeta \in \partial\Omega$, $|\zeta| < R$, $\zeta \neq 0$,

then $M^+(r)$ is a non-decreasing convex function of $\log r$ on $(0,R)$.

Proof: First suppose that $1/2 \leq \lambda < 1$. We define

$$U(z) = \begin{cases} u^+(z) , & z \in \Omega , \\ & \qquad |z| < R . \\ 0 , & z \notin \Omega , \end{cases}$$

It is clear that U is subharmonic in $\Delta(0,R)$. We know that the maximum modulus of a function subharmonic in an annulus is a convex function of $\log r$ (this is the Hadamard three-circles theorem in subharmonic form, cf. Heins, Ex. 4, p. 80). Since $M(r,u^+) = M(r,U)$, we see that Lemma 3.1 is true in this case.

If $0 < \lambda < 1/2$, we claim that on $(0,R)$

a) $M^+(r)$ is nondecreasing,

b) $M^+(r)$ is continuous on the right,

c) $M^+(r)$ is convex in $\log r$.

As a consequence, we observe that

$$(3.7) \qquad (s,t) \to (M^+(s) - M^+(t))(\log s - \log t)^{-1} , \quad 0 < t < s < R ,$$

is increasing in each variable separately. In particular, $r \frac{d}{dr} M^+(r)$ exists a.e., where the derivative is interpreted as a right hand derivative.

In the proofs of (a) - (c), we follow Lewis ([2], p. 193).

Proof of (a): Let t be given, $0 < t < 1$, and let $\Omega_t = \Delta(0,t) \cap \Omega$. Let $C = \sup u(z)$, $t \in \Omega_t$. We claim that

$$u(z) \leq \max \{\cos \pi\lambda \; C^+ , M^+(t)\} = A , \; z \in \Omega_t .$$

Indeed, if $\zeta \in \partial \Omega_t \cap \partial\Omega$ and $\zeta \neq 0$, then

$$u|_{\Omega_t}(\zeta) \leq u(\zeta) \leq \cos \pi\lambda \; M^+ (|\zeta|) \leq A ,$$

thanks to (3.6) and the fact that $\cos \pi\lambda > 0$. Moreover, since u is upper semicontinuous on Ω , we have

$$u|_{\Omega_t}(\zeta) \leq u(\zeta) \leq M^+(t) , \; \zeta \in \Omega \cap \partial\Omega_t .$$

Using these facts and the Phragmén-Lindelöf maximum principle, we see that $u(z) \leq A$, $z \in \Omega_t$. Hence if $0 < s < t < R$, then

$$M(s) \leq C \leq \max \{\cos \pi\lambda \; C^+ , M^+(t)\} ,$$

and so $M^+(s) \leq M^+(t)$. This completes the proof of a).

Proof of b): Let $s \in (0,R)$ and note that

$$M(s) \leq \lim_{r \to s+} M^+(r) = C_1 .$$

If $C_1 = 0$, then b) is trivially true. Hence we assume $C_1 > 0$. In this case, there exists a sequence $\{z_n\}_1^\infty$ where $|z_n| > s$ and $z_n \in \Omega$, $n = 1,2\ldots$, such that $u(z_n) = M(|z_n|)$, $\lim z_n = z$ and $|z| = s$ as we see from the inequality

$$u(\zeta) \leq \cos \pi\lambda\, M^+(|\zeta|) < M(|\zeta|) , \; \zeta \in \partial\Omega , \; s < |\zeta| < R .$$

Moreover, $z \in \Omega$. If this is not the case and $z \in \partial\Omega$, we know that if $\varepsilon > 0$ is given and n is large enough that

$$M(s) \leq u(z_n) \leq u(z) + \varepsilon \leq \cos \pi\lambda\, M^+(s) + \varepsilon .$$

Since $\varepsilon > 0$ is arbitrary, $M^+(s) = 0$ and it follows that $M(|z_n|) = u(z_n) \leq 0$ which contradicts the assumption $C_1 = 0$. In the inequalities, we used (3.6) and the fact that $u(z) = \limsup u(\zeta)$, $\zeta \to z$, $\zeta \in \Omega$.

Using the above facts and the upper semicontinuity of u at z, we obtain $u(z) \geq \lim_{n\to\infty} u(z_n) = C_1$. Hence $M(s) \geq C_1$. We know that $M(s) \leq C_1$, hence there is equality and b) is proved.

Proof of c): Let $0 < s < R$ and suppose that $M(s) > 0$. We assume that

$$\lim_{r \to s^-} M^+(r) > \cos \pi\lambda\, M(s) .$$

If this is false, let $z_0 \in \Omega$ be such that $u(z_0) = M(s)$. Then using b), we see that

$$M(s) = u(z_0) \leq \lim_{\delta \to 0+} (2\pi)^{-1} \int_0^{2\pi} u(z_0 + \delta\, e^{i\theta})d\theta \leq \frac{1}{2}\,(1 + \cos\pi\lambda)M(s) .$$

Since $M(s) > 0$, we have a contradiction. The assertion is proved.

To complete the proof of c), let $0 < r_1 < s < r_2 < R$ be such that $M(r_1) > \cos \pi\lambda \, M(r_2)$. Let

$$v(z) = u(z) - \cos \pi\lambda \, M(r_2) \, , \; r_1 < |z| < r_2 \, , \; z \in \Omega \, .$$

Then using (3.6) and a), we see that

$$v(\zeta) = u(\zeta) - \cos \pi\lambda \, M(r_2) \le \cos \pi\lambda \, (M(|\zeta|) - M(r_2)) \le 0 \, ,$$

$$\zeta \in \delta\Omega \, , \; r_1 < |\zeta| < r_2 \, .$$

Hence v^+ has a subharmonic extension to the annulus $\{r_1 < |z| < r_2\}$ and $M(r,v^+)$ is a convex function of log r on (r_1, r_2). But we know that

$$M(r,v) = M(r,u) - \cos \pi\lambda \, M(r_2,u) > M(r_1,u) - \cos \pi\lambda \, M(r_2,u) > 0 \, .$$

It follows that $M(r,u)$ is convex in log r and c) is proved.

<u>Lemma 3.2</u>: <u>Let u be as in Theorem 3.1</u>. <u>Then</u>

$$0 < \sup_{r>0} r^{-2\lambda} M(r) \le (\cos\pi\lambda)^{-1} \liminf_{R\to\infty} R^{-2\lambda} M(R) \, .$$

<u>Proof</u>: Consider in $\Delta(0,R) \cap \{\operatorname{Re} z > 0\}$

$$h(z) = u(z) - (\cos\pi\lambda)^{-1} R^{-2\lambda} M^+(R) \, r^{2\lambda} \cos 2\lambda\theta \, .$$

It is easy to check that

$$h(iy) \le \cos \pi\lambda \, M^+(|y|,h) \, , \; 0 < |y| < R \, , \; h(Re^{i\theta}) \le 0 \, , \; |\theta| < \pi/2 \, .$$

Applying Lemma 3.1, we see that

$$M^+(r,h) \le M^+(R,h) \le 0 \, , \; 0 < r < R \, .$$

It follows that

$$M(r) = M(r,u) \leq (\cos\pi\lambda)^{-1} R^{-2\lambda} M^+(R) r^{2\lambda} , \quad 0 < r < R ,$$

and Lemma 3.2 is proved.

As a consequence of Lemma 3.2, we note that either $r^{-2\lambda}M(r) \to \infty$, $r \to \infty$, or $M(r) = \underline{O}(r^{2\lambda})$, $r \to \infty$. In the sequel, we assume the second alternative to hold.

The next step is to construct a harmonic majorant of u. For this purpose, consider

$$v(z) = (\cos\pi\lambda/\pi) \int_{-\infty}^{\infty} x\,(x^2 + (y-t)^2)^{-1} M^+(|t|)\,dt .$$

Since $M^+(r) = \underline{O}(r^{2\lambda})$, $r \to \infty$, this integral converges to a harmonic function in $\{\mathrm{Re}\, z > 0\}$ which is such that $\lim_{r\to\infty} r^{-1} M(r,v) = 0$. Applying the Phragmén-Lindelöf theorem quoted above we see that

$$u(z) \leq v(z) , \quad \mathrm{Re}\, z > 0 . \tag{3.8}$$

Finally, we claim that

$$M(r,v) = v(r) , \quad r > 0 . \tag{3.9}$$

To prove (3.9), we observe that

$$v(re^{i\theta}) = \int_0^{\infty} A(\theta,y)\, M^+(ry)\,dy , \tag{3.10}$$

where $A(\theta,y) = (\cos \pi\lambda/\pi) \cos\theta\{(1-2y \sin\theta+y^2)^{-1} + (1+2y \sin\theta+y^2)^{-1}\}$.

Since the function defined in (3.7) is increasing in each variable separately, it follows by monotone convergence that

$$r \frac{\partial}{\partial r} v(re^{i\theta}) = \int_0^{\infty} A(\theta,y)\,(ry) \frac{d}{d(ry)} M^+(ry)\,dy .$$

By the convexity properties of $M^+(r)$ discussed in Lemma 3.1, $r \frac{d}{dr} M^+(r)$ and thus also $r \frac{\partial}{\partial r} v(re^{i\theta})$ is a nondecreasing function on $(0,\infty)$. Hence

$$\frac{\partial}{\partial r} \left(r \frac{\partial}{\partial r} v(re^{i\theta})\right) \geq 0 \;, \; \text{Re } z > 0 \;.$$

Since v is harmonic in $\{\text{Re } z > 0\}$, this derivative exists in $\{\text{Re } z > 0\}$ and it follows that $v_{\theta\theta}(re^{i\theta}) \leq 0$, $|\theta| < \pi/2$. We have here expressed the Laplace operator in polar coordinates. But $v(re^{i\theta}) = v(re^{-i\theta})$, $|\theta| \leq \pi/2$, and it is now clear that $\sup v(re^{i\theta}) = v(r)$, $|\theta| < \pi/2$. This proves (3.9). Thus

$$M^+(r) \leq M^+(r,v) = v(r) = (2 \cos \pi\lambda/\pi) \int_0^\infty r(r^2 + t^2)^{-1} M^+(t)\, dt \;,$$

$$r^{-2\lambda} M^+(r) \leq (2 \cos \pi\lambda/\pi) \int_0^\infty (r/t)^{1-2\lambda}(1 + r^2/t^2)^{-1} t^{-2\lambda} M^+(t)\, dt/t \;.$$

This is a convolution inequality of exactly the same type as the one discussed in § 2. It follows that

$$\lim_{r\to\infty} r^{-2\lambda} M^+(r) = \alpha > 0 \;. \tag{3.11}$$

$$\int_1^\infty t^{-1-2\lambda}(v(t) - M^+(t))dt < \infty \;. \tag{3.12}$$

The fact that $\alpha > 0$ is a consequence of (3.3) and Lemma 3.2.

By (3.11) and the Phragmén-Lindelöf theorem quoted above, it follows that

$$\lim \,(r^{-2\lambda} v(re^{i\theta}) - \alpha \cos 2\lambda\theta) = 0 \;, \; r \to \infty \;, \tag{3.13}$$

uniformly in θ, $|\theta| \leq \pi/2$.

We now know the behavior of v at ∞ ; it remains to study $p = v-u$ which is a nonnegative, superharmonic function in $\{\text{Re } z > 0\}$. If r is given, there exists z_r , $|z_r| = r$, such that $u(z_r) = M(r)$. Hence

$$0 \le p(z_r) = v(z_r) - u(z_r) \le v(r) - M(r) . \tag{3.14}$$

By (3.11) and (3.13) ,

$$\lim_{r\to\infty} r^{-2\lambda} p(z_r) = 0 . \tag{3.15}$$

Furthermore, by (3.13) and (3.14) ,

$$\lim_{r\to\infty} \arg z_r = 0 \tag{3.16}$$

Let δ be a given small positive number. We introduce $w(r) = \inf p(re^{i\theta})$, $|\theta| < \delta$. Using (3.12), (3.14) and (3.16), we see that

$$\int_1^\infty r^{-1-2\lambda} w(r)\, dr < \infty . \tag{3.17}$$

The convergence of this integral is what we need to study the behavior of p at ∞ . From the Riesz representation theorem and the Phragmén-Lindelöf theorem (here (3.15) is important!), we see that

$$p(z) = \int_{-\infty}^{\infty} x(x^2 + (y-t)^2)^{-1}\, d\gamma(t) + \int_{\mathrm{Re}\,\zeta>0} g(z,\zeta)\, d\mu(\zeta) .$$

Here γ and μ are (nonnegative) measures and $g(z,\zeta) = \log|(z + \bar{\zeta})/(z - \zeta)|$ is Green's function for $\{\mathrm{Re}\, z > 0\}$. We claim that

$$\int_{|t|<1} d\gamma(t) + \int_{|t|\ge 1} |t|^{-1-2\lambda}\, d\gamma(t) < \infty , \tag{3.18}$$

$$\int_{|\zeta|<1} \xi\, d\mu(\zeta) + \int_{|\zeta|\ge 1} \xi|\zeta|^{-1-2\lambda}\, d\mu(\zeta) < \infty , \tag{3.19}$$

where $\zeta = \xi + i\eta$, $\xi,\eta \in \underline{R}$.

Proof of (3.19): Let $z = re^{i\theta}$ and $\zeta = \rho e^{i\varphi}$. We note that if $\rho > 2r$, or if $\rho < r/2$,

$$g(z,\zeta) \approx 2\rho r \cos\theta \cos\varphi \, (\rho - r)^{-2} .$$

Using this estimate, (3.17) and Harnack's inequality, we see that

$$\infty > \int_1^\infty r^{-1-2\lambda} \inf_{|\theta| < \delta} \int_{|\zeta| > 2r} g(z,\zeta)\, d\mu(\zeta)\, dr \geq$$

$$\text{Const.} \int_1^\infty r^{-1-2\lambda} \int_{|\zeta|>2r} r\xi(\rho-r)^{-2}\, d\mu(\zeta)\, dr \geq$$

$$\geq \text{Const.} \int_{|\zeta| > 2}^{\infty} |\zeta|^{-1-2\lambda}\, \xi \, d\mu(\zeta) .$$

To estimate the integral over $\Delta(0,1) \cap \{\mathrm{Re}\, \zeta > 0\}$, we replace integration over $\{|\zeta| > 2r\}$ in the formula above by integration over $\{|\zeta| < r/2\}$.

To prove (3.18), we use (3.17) once more. By the same argument as above,

$$\infty > \int_1^\infty r^{-1-2\lambda} \inf_{|\theta|<\delta} \int_{-\infty}^{\infty} x\,(x^2 + (y-t)^2)^{-1}\, d\gamma(t) \geq$$

$$\geq \text{Const.} \int_1^\infty r^{-1-2\lambda} \int_{-\infty}^{\infty} r(r^2 + t^2)^{-1}\, d\gamma(t) \geq$$

$$\geq \text{Const.} \{ \int_{|t|<1} d\gamma(t) + \int_{|t|\geq 1} |t|^{-1-2\lambda}\, d\gamma(t)\} .$$

Thus (3.18) is true.

We now apply the technique of Hayman [1] and Azarin [1]. We introduce, if $z = x + iy, x, y \in \underline{R}$,

$$d\nu(\zeta) = \begin{cases} 2\xi(1 + |\zeta|^{1+2\lambda})^{-1}\, d\mu(\zeta) , & \xi > 0 , \\ (1 + |\eta|^{1+2\lambda})^{-1}\, d\gamma(\eta) , & \xi = 0 , \end{cases}$$

$$K(z,\zeta) = \begin{cases} (1 + |\zeta|^{1+2\lambda})\, g(z,\zeta)(2\xi)^{-1} , & \xi > 0 , \\ x(1 + |\eta|^{1+2\lambda})\, |z - i\eta|^{-2} , & \xi = 0 . \end{cases}$$

In this notation, if $D = \{\mathrm{Re}\, z \geq 0\}$, $p(z) = \int_D K(z,\zeta)\, d\nu(\zeta)$.

It follows from (3.18) and (3.19) that

$$\int_D d\nu(\zeta) < \infty . \tag{3.20}$$

In the case $\lambda = 1/2$, these formulas can be found in Hayman [1]. Let $D = D_1 \cup D_2 \cup D_3$, where

$$D_1 = D \cap \{\zeta |\ |\zeta| \leq |z|/2\} ,$$

$$D_2 = D \cap \{\zeta |\ |z|/2 < |\zeta| < 3|z|/2\} ,$$

$$D_3 = D \cap \{\zeta |\ |\zeta| \geq 3|z|/2\} .$$

It is easily seen that

$$g(z,\zeta) \leq \text{Const. } x\xi|z|^{-2} , \quad \zeta \in D_1 ,$$

$$g(z,\zeta) \leq \text{Const. } x\xi|\zeta|^{-2} , \quad \zeta \in D_3 .$$

Let $\int_{D_k} K(z,\zeta)\, d\nu(\zeta) = I_k(z)$, $k = 1,2,3$.

By (3.18) and (3.19), $\lim r^{-2\lambda}(I_1 + I_3)(z) = 0$, uniformly in θ, $|\theta| \leq \pi/2$.

Let us give the details for I_3 . If $|\zeta| > 3|z|/2 > 1$, we see that

$$K(z,\zeta) \leq \text{Const. } x|\zeta|^{-1+2\lambda} \ , \ \xi \geq 0 \ .$$

Thus, recalling that $0 < \lambda < 1/2$, we obtain

$$I_3(z) = \int_{D_3} K(z,\zeta)d\nu(\zeta) \leq \text{Const. } x|z|^{-1+2\lambda} \int_{D_3} d\nu(\zeta)$$

and thus

$$\lim_{z\to\infty} r^{-2\lambda} I_3(z) \leq \text{Const. } \lim_{z\to\infty} \int_{D_3} d\nu(\zeta) = 0 \ .$$

It remains to consider $I_2(z)$. We need the following concept (cf. Hayman [1]):

_If_ $\varepsilon > 0$ _is given, we say the z is_ ε_-normal if for_ $0 < h \leq |z|/2$,

$$\int_{D(z)} d\nu(\zeta) < \varepsilon h/r \ ,$$

where $D(z) = D \cap \{\zeta |\ |\zeta - z| < h\}$.

Lemma 3.3: _If z is_ ε_-normal_, $|z| > 1$, _then_

$$|I_2(z)| \leq \text{Const. } |z|^{2\lambda}(\varepsilon + \int_{D(z)} d\nu(\zeta)) \ .$$

Remark: The case $\lambda = 1/2$ is in Hayman [1] (and in Azarin [1] in $\underline{R}^d$, $d \geq 2$). The same technique also works in the situation considered here. A similar discussion in $\underline{R}^d$, $d \geq 3$, can be found in the proof of Lemma 4 in Essén-Lewis [1].

Proof of Lemma 3.3: Let

$$\Omega_n = \{\zeta \in D_2 \colon 2^{n-1}\, x \leq |z-\zeta| < 2^n\, x\} \ , \ n \in \underline{Z} \ .$$

We suppose that $|z| > 1$ and that z is ε-normal so that $\nu(z) = 0$. We want an estimate of

$$\int_{D_2} K(z,\zeta)\, d\nu(\zeta) = \sum_{-\infty}^{\infty} \int_{\Omega_n} K(z,\zeta)\, d\nu(\zeta) = \sum_{-\infty}^{\infty} J_n(z) .$$

Let us first consider the case $n \leq -1$. Then, if $\zeta \in \Omega_n$, the distance from ζ to ∂D is at least $x/2$ and $|z + \bar{\zeta}| < 3x$. We have the following estimate:

$$K(z,\zeta) \leq \text{Const. } x^{-1} |z|^{1+2\lambda} \log (3/2^{n-1}) , \ \zeta \in \Omega_n .$$

If $h = 2^n x \leq |z|/2$, we know by assumption that

$$\text{(3.21)} \qquad \int_{D(z)} d\nu(\zeta) \leq \varepsilon\, h/|z| \leq \varepsilon\, 2^n x/|z| ,$$

and we obtain

$$\text{(3.22)} \qquad J_n(z) \leq \text{Const. } \varepsilon\, |z|^{2\lambda}\, 2^n \log (3/2^{n-1}) , \ n \leq -1 .$$

If $n > 0$, we use the estimate

$$g(z,\zeta) \leq 4\, x\, \xi\, |z - \zeta|^{-2}$$

which gives us

$$K(z,\zeta) \leq \text{Const. } x\, |z|^{1+2\lambda}\, |z - \zeta|^{-2} .$$

If $h = 2^n x \leq |z|/2$, (3.21) is still correct and we obtain

$$\text{(3.23)} \qquad J_n(z) \leq \text{Const. } \varepsilon\, |z|^{2\lambda}\, 2^{-n} .$$

If $h = 2^n x > |z|/2$, we obtain

$$\text{(3.24)} \qquad J_n(z) \leq \text{Const. } x|z|^{1+2\lambda} \int_{\Omega_n} d\nu(\zeta)\, 2^{-2n} x^{-2} \leq$$

$$\leq \text{Const. } |z|^{2\lambda}\, 2^{-n} \int_{D_2} d\nu(\zeta)$$

Combining (3.22) - (3.24), we see that if z is ε-normal and $|z| \geq 1$,

$$\int_{D_2} K(z,\zeta)\, d\nu(\zeta) \leq \text{Const. } |z|^{2\lambda}\{\varepsilon + \int_{D_2} d\nu(\zeta)\} .$$

Adding up, we know that if z is ε-normal and $|z|$ is large enough, we have the estimate

$$(I_1 + I_2 + I_3)(z) \leq \text{Const. } |z|^{2\lambda}\, \varepsilon .$$

To conclude the proof of Theorem 3.1, we need the following lemma of Azarin which in its turn depends on a lemma of Landkof (cf. Lemma 3.2, p. 197 in Landkof).

Lemma 3.4: The set $\Delta(\varepsilon)$ of points not ε-normal may be covered by a system $F(\varepsilon)$ of disks $\{B_i\}$ whose radii $\{r_i\}$ and distances $\{R_i\}$ from their centres to the origin satsfy (3.5).

Now choose an increasing sequence $\{t_n\}$ of positive numbers, $t_n \to \infty$ such that for $B_i \in F(1/n)$,

$$\sum_{R_i > t_n} r_i/R_i < 2^{-n} .$$

If $F(1/n,t_n)$ is the set of disks whose radii appear in this sum, we put $F_0 = \bigcup_1^\infty F(1/n,t_n)$. It is clear that

$$I_2(z) \leq \text{Const. } n^{-1} |z|^{2\lambda} , \quad |z| > t_n , \quad z \in D \setminus F_0 .$$

This shows that $p = I_1 + I_2 + I_3$ has the right kind of behavior at infinity. Since we know the behavior of v, Theorem 3.1 is proved for $u = v-p$.

4. The Paley conjecture

Let u be subharmonic in $\underline{C}$ of lower order λ_0. Let

$$T(r,u) = T(r) = (2\pi)^{-1} \int_0^{2\pi} u^+(re^{i\theta})\, d\theta .$$

What can be said about the relation between $\limsup_{r\to\infty} T(r)/M(r)$ and λ_0 ? The result given below in $\underline{C}$ is known as the Paley conjecture. In 1969, Petrenko proved an analogous result for meromorphic functions. An account of Petrenko's proof is given in Fuchs [1]. The problem for subharmonic functions in $\underline{R}^d$, $d \geq 3$, was solved by B. Dahlberg (cf. [1], where further references are given). In this section, we shall essentially give Dahlberg's proof in the case $d = 2$. The main result is

Theorem 4.1:

(4.1) $$\limsup_{r\to\infty} T(r)/M(r) \geq \begin{cases} \sin \pi\lambda_o/\lambda_o\pi , & 0 < \lambda_o \leq 1/2 , \\ (\pi\lambda_0)^{-1} , & \lambda_0 \geq 1/2 . \end{cases}$$

In the sequel, we assume for simplicity that $\lambda_0 \geq 1/2$. We leave the modifications which are necessary when $0 < \lambda_0 < 1/2$ to the reader (see also § 5 in Dahlberg [1]).

Theorem 4.1 is a consequence of

Theorem 4.2: *Let u be subharmonic and non-constant in $\underline{C}$. Let λ be given, $1/2 \leq \lambda < \infty$. If*

(4.2) $$T(r) \leq (\pi\lambda)^{-1} M(r) , \quad r > r_0 ,$$

then $\alpha = \lim_{r\to\infty} r^{-\lambda} M(r)$ exists and $\alpha > 0$. The case $\alpha = \infty$ can occur.

Let us first give the proof of Theorem 4.1 in the case $\lambda_0 \geq 1/2$ assuming Theorem 4.2 to be true. If (4.1) is false, there exist numbers λ and r_0, $\lambda > \lambda_0$, such that (4.2) is true. By Theorem 4.2, $\lim_{r\to\infty} r^{-\lambda}M(r) > 0$. This contradicts the fact that $\liminf_{r\to\infty} \log M(r)/\log r = \lambda_0 < \lambda$, and the proof is complete.

Theorem 4.1 is best possible. In fact, there is equality in (4.1) for the function

$$(4.3) \qquad u^+(re^{i\theta}) = \begin{cases} r^\lambda \cos \lambda\theta \ , & |\theta| < \pi/2\lambda \ , \\ 0 \ , & |\theta| > \pi/2\lambda \ , \end{cases}$$

if we take $\lambda = \lambda_0 \geq 1/2$.

Proof of Theorem 4.2: Let $K = \{z: |\arg z| < \pi/2\lambda\}$. Let $w = \rho e^{i\theta}$. The Neumann function of K with pole at (r,0) is given by

$$N(r,w) = -(1/2)\{\log|1-(w/r)^{2\lambda}| + \log|1-(\bar{w}/r)^{2\lambda}|\} =$$

$$= -(1/2)\log |1 - 2 (\rho/r)^{2\lambda} \cos 2\lambda\theta + (\rho/r)^{4\lambda}| \ .$$

Exercize: The Neumann function N(z,w) can be written as an infinite series containing the normalized eigenfunctions of a certain boundary problem which we find if we separate the variables in Laplace's equation in K, assuming the normal derivative of the solution to be zero on ∂K. Details in $\underline{R}^d$, $d \geq 3$, can be found in Dahlberg [1]. Use this information to deduce the expressions given above for N(r,w) !

We also introduce, for r and ρ positive,

$$\Psi(r,\rho) = N(r,\rho)\exp\{\pm i\pi/2\lambda\}) = -\log\ (1 + (\rho/r)^{2\lambda}) \ .$$

We define, for $w = \rho e^{i\theta}$,

$$\Psi(r,w) = \Psi(r,|w|) = \Psi(r,\rho) \ .$$

The Neumann function has the following properties.

(4.4) $$\frac{\partial}{\partial n} N(r,w) = 0 , \qquad w \in \partial K .$$

(4.5) $$N(r,w) + \log |r-w| \text{ is harmonic, } w \in K .$$

(4.6) $$N(r,w) \geq \Psi (r,\rho) , \qquad w \in K .$$

(4.7) $$\Psi(r,\cdot) \text{ is superharmonic in } \underline{C} .$$

Proof of (4.7): $-\Psi(r,w)$ is the maximum over a circle of radius $|w|$ of a subharmonic function.

(4.8) $$P(r,\rho) = -\rho\Delta\Psi(r,w) = 4\lambda^2 \rho^{-1}((r/\rho)^{\lambda} + (\rho/r)^{\lambda})^{-2} .$$

Let $K_R = K \cap \Delta(0,R)$ and $D_R = K \cap C(0,R)$. The basic tool in the proof of Theorem 4.2 is

Lemma 4.1: Let u be subharmonic in C, let $u \in C^{(2)}$ and assume that $\Delta u = 0$ in $\Delta(0,1)$. Then, if r,R are given, $0 < r < R/2$,

(4.9) $$u(r) \leq V(u,r,R) + \text{Const. } M(6R,u)(r/R)^{2\lambda}$$

where

$$V(u,r,R) = -(2\pi)^{-1} \iint_{K_R} u(z)\, \Delta\Psi(r,z)\, dA(z) .$$

Here dA(z) denotes Lebesgue measure in the plane.

We first give the proof of Theorem 4.2, assuming Lemma 4.1 to hold. Assume $M(r_0,u) > 0$ and form $v = (u - M(r_0,u))^+$ which has the following properties:

(4.10) $$v \geq 0 ,$$

(4.11) $$v(z) = 0 , \ |z| < r_0 ,$$

(4.12) $$M(r,v) = M(r,u) - M(r_0,u) \ , \quad r > r_0 \ ,$$

(4.13) $$T(r,v) \leq (\pi\lambda)^{-1} M(r,v) \ , \quad r > 0 \ .$$

Proof of (4.13): Let, if r is given, $\Omega = \{\theta : u^+(re^{i\theta}) \geq M(r_0,u)\}$ and put $\omega = \int_\Omega d\theta$. If $\omega \leq 2\lambda^{-1}$,

$$T(r,v) = (2\pi)^{-1} \int_\Omega (u(re^{i\theta}) - M(r_0,u))^+ \, d\theta \leq (2\pi)^{-1} \, 2\lambda^{-1} \, M(r,v) =$$

$$= (\pi\lambda)^{-1} M(r,v) \ .$$

If $\omega > 2\lambda^{-1}$ and $r > r_0$, then by (4.2)

$$T(r,v) = (2\pi)^{-1} \int_\Omega (u^+(re^{i\theta}) - M(r_0,u)) \, d\theta \leq T(r,u^+) - (\omega/2\pi) \, M(r_0,u) \leq$$

$$\leq (\pi\lambda)^{-1} (M(r,u^+) - M(r_0,u)) = (\pi\lambda)^{-1} M(r,v) \ ,$$

and (4.13) is proved.

We first assume that $u \in C^{(2)}$ so that we can use Lemma 4.1. If r is given, we can assume that $M(r,v) = v(r)$, since rotation of the coordinate system does not change any of our basic assumptions.

Hence if $r < R/2$,

(4.14) $$M(r,v) = v(r) \leq V(v,r,R) + \text{Const.} \ M(6R,v)(r/R)^{2\lambda} \ .$$

If $v \notin C^{(2)}$, we choose a decreasing sequence $\{v_n\}$ in $C^{(2)}$ converging to v . Since (4.14) holds for each v_n , it will also be true for v . Using (4.8), we see that

(4.15) $$M(r,v) \leq \int_0^R P(r,\rho) \, T(\rho,v) \, d\rho + \text{Const.} \ M(6R,v)(r/R)^{2\lambda} \ .$$

If we apply Green's formula to the function defined in (4.3) and to $\Psi(r,\cdot) - N(r,\cdot)$ in the angular domain K, we see that

(4.16) $$r^{\lambda} = (\pi\lambda)^{-1} \int_0^{\infty} P(r,\rho)\, \rho^{\lambda}\, d\rho \ ,$$

(for details, see Lemma 4.2 below).

We have now a situation of exactly the same type as in the proof of the cos $\pi\lambda$-theorem. Using (4.13), we deduce an inequality for $r^{-\lambda} M(r,v)$ from (4.15). From (4.16), we see that the integral of the positive kernel over $(0,\infty)$ is 1. The simple argument at the end of § 1 now shows that

$$r^{-\lambda} M(r,v) \leq \text{Const. } R^{-\lambda} M(R,v) \ , \ 0 < r < R \ .$$

Thus either $r^{-\lambda} M(r,v) \to \infty$, or $M(r,v) = \underline{O}(r^{\lambda})$, $r \to \infty$. Applying Theorem 2.2, we obtain the existence of the positive limit $\alpha = \lim_{r\to\infty} r^{-\lambda} M(r,v)$. This concludes the proof of Theorem 4.2.

It remains to prove Lemma 4.1. We introduce $\varepsilon(r,z) = \Psi(r,z) - N(r,z)$, $z \in K$. The following lemma is proved using the known expressions for Ψ and N. Let $d(z) = \text{dist}(z, \partial K)$, where ∂K is the boundary of K.

<u>Lemma 4.2</u>: <u>If</u> $w = \rho e^{i\theta} \in K$ <u>and</u> $\rho > 2r$,

(4.17) $$-\text{Const. } d(w)^2\, r^{2\lambda}\, |w|^{-2\lambda-2} < \varepsilon(r,w) < 0 \ ,$$

(4.18) $$\left|\frac{\partial}{\partial\rho} \varepsilon(r,\rho e^{i\theta})\right| < \text{Const. } (r/\rho)^{2\lambda}\, d(w)^2\, \rho^{-3} \ .$$

The basic idea in the proof of Lemma 4.1 is to use Green's identity in $K_R = K \cap \Delta(0,R)$ in such a way that there will be no contribution in the final formula from the rays $\{\rho \exp(\pm i\, \pi/2\lambda)\}$, $\rho > 0$. In this context, it is natural to introduce $\varepsilon(r,\cdot)$, since $\varepsilon(r,w) = \frac{\partial}{\partial n} \varepsilon(r,w) = 0$ on these two rays. Applying Green's formula to the functions $g \in C^{(2)}$ and $\varepsilon(r,\cdot)$ in K_R, we see that

$$(4.19)\qquad g(r) = V(g,r,R) + (2\pi)^{-1} \iint_{K_R} \varepsilon(r,w)\Delta g(w)dA(w) + S(g,r,R)$$

where

$$S(g,r,R) = (2\pi)^{-1} \int_{D_R} \left(g(w) \frac{\partial}{\partial n} \varepsilon(r,w) - \varepsilon(r,w) \frac{\partial}{\partial n} g(w)\right) ds ,$$

(ds is the element of arclength on the arc D_R).

Taking $g = u$, where u is as in Lemma 4.1, we see that

$$(4.20)\qquad u(r) = V(u,r,R) + (2\pi)^{-1} \iint_{K_R} \varepsilon(r,z)\Delta u(z)dA(z) + S(u,r,R) .$$

Let h be the least harmonic majorant of u in K_{3R} . If $G_{3R}(z,\cdot)$ is Green's function in K_{3R} with pole at z, we know that $u = h-p$, where

$$p(z) = (2\pi)^{-1} \iint_{K_{3R}} G_{3R}(z,w)\ \Delta u(w)\ dA(w) ,$$

$$S(u,r,R) = S(h,r,R) - S(p,r,R) .$$

The following estimate is well known:

Let Max $|h(z)| = M$, $|z| \in K_{2R}$. Then

$$(4.21)\qquad |\text{grad } h(z)| \leq \text{Const. } d(z)^{-1} M , \quad z \in D_R .$$

Applying Lemma 4.2 we conclude that

$$|S(h,r,R)| \leq \text{Const. } M(3R)(r/R)^{2\lambda} , \quad r < R/2 .$$

Comparing (4.9) and (4.20), we see that the remaining problem is to estimate

$$H(r) = (2\pi)^{-1} \iint_{K_R} \Delta u(z)\ \varepsilon(r,z)\ dA(z) - S(p,r,R) .$$

We note that $\Delta p = -\Delta u$. From (4.19) with $g = p$, we see that

$$p(r) = V(p,r,R) - (2\pi)^{-1} \iint_{K_R} \varepsilon(r,z)\, \Delta u(z)\, dA(z) + S(p,r,R) ,$$

and thus that

$$H(r) = V(p,r,R) - p(r) = (2\pi)^{-1} \iint_{K_{3R}} E(r,R,w)\, \Delta u(w)\, dA(w) , \tag{4.22}$$

where

$$E(r,R,w) = G_{3R}(r,w) + (2\pi)^{-1} \iint_{K_R} G_{3R}(z,w)\, \Delta \Psi(r,z)\, dA(z) .$$

In the last step, we changed the order of integration in an absolutely convergent integral.

<u>Lemma 4.3</u>: $E(r,R,w) = G_{3R}(r,w) - V(G_{3R}(\cdot,w),r,R) \geq$

$$\geq -\text{Const.}\ (r/R)^{2\lambda} , \quad 0 < r < R/2 , \quad w \in K_{3R} .$$

Assuming Lemma 4.3 to hold, we see from (4.22) that

$$H(r) \leq \text{Const.}\ (r/R)^{2\lambda} \iint_{|w| < 3R} \Delta u(w)\, dA(w) . \tag{4.23}$$

We remind the reader that u was assumed to be subharmonic in $\underline{C}$. It is well-known that for all sufficiently large values of R,

$$\iint_{|w| < 3R} \Delta u(w)\, dA(w) \leq \text{Const.}\ M(6R,u) . \tag{4.24}$$

To prove (4.24), we consider $I(r) = \int_0^{2\pi} u(re^{i\theta})\, d\theta$. It is known that $I(r)$ is a convex function of $\log r$, (cf. Heins, Ex. 4, p. 80). Let $I'(r)$ be the right hand derivative of $I(r)$ at r. Then

$$rI'(r) \leq (I(2r) - I(r))/(\log 2r - \log r) \leq (\log 2)^{-1} \{I(2r) - 2\pi\, u(0)\}$$

In the last step, we used the subharmonic mean value property.

Applying Green's formula, we obtain

$$\iint_{|w| < r} \Delta u(w)\, dA(w) = \int_0^{2\pi} r \frac{\partial}{\partial r} (u(re^{i\theta}))d\theta = rI'(r) \leq$$

$$\leq (\log 2)^{-1} (I(2r) - 2\pi u(0)) \leq (\log 2)^{-1}(M(2r) - 2\pi u(0)) .$$

Putting $r = 3R$, we obtain (4.24) (for an argument using Jensen's formula, cf. Heins [1], p. 202).

Our basic estimate (4.9) is now a consequence of (4.23) and (4.24).

It remains to prove Lemma 4.3. For this purpose, we first prove

Lemma 4.4: *Let $G(\cdot,w)$ be Green's function in K. Then if $r > 0$,*

$$G(r,w) \geq V(G(\cdot,w),r,R) .$$

Proof: Consider

$$F(r,w) = \varepsilon(r,w) + G(r,w) = \Psi(r,w) - N(r,w) + G(r,w) .$$

We note that

a) $F(r,\cdot)$ is superharmonic in K and $\Delta F(r,w) = \Delta\Psi(r,w)$.

b) $F(r,w) = 0$, $w \in \partial K$,

c) $F(r,w) \to 0$, $w \to \infty$ in K .

Thus the greatest harmonic minorant of $F(r,\cdot)$ is zero and $F(r,\cdot)$ must be a potential (cf. Helms, Corollary 6.19), and we obtain

$$G(r,w) \geq F(r,w) = -(2\pi)^{-1} \iint_K G(z,w)\, \Delta\Psi(r,w)\, dA(z) \geq V(G(\cdot,w),r,R) .$$

The proof of Lemma 4.4 is complete.

To prove Lemma 4.3, let $w,z \in K_{3R}$. Let $\tilde{z}$ be the reflection of z in C_{3R}, e.g., $\tilde{z} = z(3R)^2|z|^{-2}$. It is well-known that

$$G_{3R}(z,w) = G(z,w) - G(\tilde{z},w) = G(z,w) - \Phi(z,w) .$$

Hence

$$E(r,R,w) = G(r,w) - \Phi(r,w) - V(G(\cdot,w),r,R) + V(\Phi(\cdot,w),r,R) . \tag{4.25}$$

Applying (4.19) with $g = \Phi(\cdot,w)$ which is harmonic in K_R, we see that

$$\Phi(r,w) = V(\Phi(\cdot,w),r,R) + S(\Phi(\cdot,w),r,R) .$$

Using this formula and Lemma 4.4, it follows from (4.25) that

$$E(r,R,w) \geq -S(\Phi(\cdot,w),r,R) . \tag{4.26}$$

The right hand member is an integral over D_R and there is no singularity of $\Phi(\cdot,w)$ near D_R when $w \in K_{3R}$. We note that

$$\sup_{z\in K_{2R}} \Phi(z,w) = \sup_{z\in D_{2R}} \Phi(z,w) = \sup_{z\in D_{9R/2}} G(z,w) \leq \text{Const.} , \quad w \in K_{3R} . \tag{4.27}$$

This constant is independent of R since $G(z,w) = G(z/R, w/R)$.

Applying (4.21), we see from (4.27) and Lemma 4.2 that

$$|S(\Phi(\cdot,w),r,R)| \leq \text{Const.}\ (r/R)^{2\lambda}$$

From (4.26), we now obtain Lemma 4.3 and the proof is complete.

5. A cos $\pi\lambda$-problem and a differential inequality

Let u be a non-constant function which is subharmonic in the complex plane $\underline{C}$. Let ψ: $[0,\infty) \to [0,1)$ be a lower semicontinuous function such that

(5.1) $$m(r) \leq \cos \pi\psi(r)\, M(r) \;, \quad r > 0 \;.$$

What can we say about the growth of M(r) as $r \to \infty$? It has been conjectured by B. Kjellberg that if (5.1) holds,

(5.2) $$M(r) \leq \text{Const.}\; M(R) \exp \{- \int_r^R \psi(t)/t \; dt\} \;, \quad 0 < r < R \;.$$

This conjecture is known to be true in the following special cases. Let λ be given, $0 < \lambda < 1$.

i) $\psi(r) = \lambda$, $r > 0$. Then (5.2) is essentially the corollary of the Hellsten-Kjellberg-Norstad inequality discussed in § 1.

ii) Let E be a union of intervals in [0,R] and assume that

(5.3) $$m(r) \leq \cos \pi\lambda \; M(r) \;, \quad r \in E \;.$$

This means that (5.1) holds with

$$\psi(r) = \begin{cases} \lambda \;, & r \in E \\ 0 \;, & r \in [0,R] \setminus E \;. \end{cases}$$

The case $\lambda = 1/2$ is due to A. Beurling (1933). The case $0 < \lambda < 1$ is in J. Lewis [1].

Unfortunately, it does not always follow from (5.1) that (5.2) is true. A counterexample is given in § 6.

We shall prove the following result. In this context, it is no essential restriction to assume that u is harmonic in $\Delta(0,1)$ so that $u(0)$ is finite. We can also assume that $u(0) \leq 0$. (If this does not hold, replace u by $u - u^+(0)$. Our basic assumption (5.1) is true also for the new function.)

Theorem 5.1: Let u and ψ be as above and assume further that ψ is a step-function taking finitely many values. Let the smallest positive number in the range of ψ be μ . Let

$$\Psi(r) = \mu\{(1 - \cos \pi\psi(r))/(1 - \cos \pi\mu)\}^{1/2} .$$

If (5.1) is true, there exists an absolute constant (which does not depend on μ) and a number r_0 only depending on u such that

$$(5.4) \quad M(r) \leq \text{Const.} \; (|u(0)| + M(R) \exp \{- \int_r^R \Psi(t) \, dt/t\}) \; , \; r_0 < r < R .$$

Corollary 1: If the range of ψ is the set $\{0,\lambda\}$, where $0 < \lambda < 1$, then

$$M(r) \leq \text{Const.} \; M(R) \exp \{-\lambda m_\ell E(r,R)\} \; , \; r_0 < r < R .$$

where E is defined in (5.3) and $m_\ell E(r,R) = \int_{E\cap(r,R)} dt/t$.

This is the theorem of Lewis quoted above. As a second application, we discuss results of P.D. Barry [1] (see also W.K. Hayman [2]). Let the order and the lower order of u be ρ_0 and λ_0 , respectively. Let

$$F_\rho = \{r\colon m(r) \leq \cos \pi\rho \; M(r)\} \; ,$$

$$\overline{\Lambda} \; F_\rho = \limsup_{R\to\infty} m_\ell \; F_\rho(1,R)/\log R \; ,$$

$$\underline{\Lambda} \; F_\rho = \liminf_{R\to\infty} m_\ell \; F_\rho(1,R)/\log R \; .$$

Corollary 2: *Let* $0 < \rho_0 < 1$. *For* $\rho_0 < \rho < 1$,

(5.5) *either* $\overline{\Lambda}\, F_\rho = 0$ *or* $\underline{\Lambda}\, F_{\rho_0} < 1$,

(5.6) *either* $\underline{\Lambda}\, F_\rho = 0$ *or* $\overline{\Lambda}\, F_{\rho_0} < 1$.

If there exists a set $E(\rho_0)$ *such that* $\overline{\Lambda}\, E(\rho_0) = 0$ *and*

(5.7) $$\limsup m(r)/M(r) \leq \cos \pi\rho_0 \ , \ r \to \infty \ , \ r \notin E(\rho_0) \ ,$$

then $\overline{\Lambda}\, F_\rho = 0$, $\rho_0 < \rho < 1$, *and there exists a set* $G(\rho_0)$, $\overline{\Lambda}\, G(\rho_0) = 0$ *such that*

(5.8) $$m(r)/M(r) \to \cos \pi\rho_0 \ , \ r \to \infty \ , \ r \notin G(\rho_0) \ .$$

Corollary 3: *Let* $0 < \lambda_0 < 1$. *If there exists a set* $E(\lambda_0)$ *such that* $\overline{\Lambda}\, E(\lambda_0) = 0$ *and*

(5.9) $$\limsup m(r)/M(r) \leq \cos \pi\lambda_0 \ , \ r \to \infty \ , \ r \notin E(\lambda_0) \ ,$$

then $\underline{\Lambda}\, F_\rho = 0$, $\lambda_0 < \rho < 1$.

Remark 1: Since $\Psi(r) < \psi(r)$ when $\mu < \psi(r) < 1$, (5.4) is slightly weaker than the Kjellberg conjecture. We note that if $\psi(r) - \mu$ is small

$$\Psi(r) \approx \mu + \frac{\pi\mu}{2} \cot \frac{\pi\mu}{2} \, (\psi(r) - \mu) \ .$$

The coefficient for $(\psi(r) - \mu)$ decreases from 1 to 0 when $0 \leq \mu \leq 1$. This means that as μ increases, the distance to the estimate conjectured by Kjellberg will also increase.

Remark 2: If ψ is non-decreasing, it can be shown that the conjecture is true

Using a different method, we have proved the following result (cf. Essén [3]). Let λ and ρ be given, $0 < \lambda < 1$ and let $\rho - \lambda$ be small compared to ρ. Let E be a union of intervals on $(0,\infty)$ and let $F = (0,\infty)\setminus E$. If

$$m(r) \leq \begin{cases} \cos \pi\lambda \, M(r) \; , & r \in E \; , \\ \cos \pi\rho \, M(r) \; , & r \in F \; , \end{cases}$$

then

$$M(r) \leq \text{Const.}M(R)\exp\{-\lambda m_\ell E(r,R) - \rho m_\ell F(r,R) + b(\rho)(\rho-\lambda)^2 m_\ell E(r,R)\}, 1<r<R,$$

where $b(\rho) = \frac{1}{2}\{\rho(1-\rho)^{-1} - \log(1-\rho)\}$.

If $\rho - \lambda$ is small, and if the range of ψ is $\{\rho,\lambda\}$, this result is sharper than the one given in Theorem 5.1. This later result, however, holds in a more general situation.

It is clear that the restriction on the range of ψ in Theorem 5.1 is unessential. If the theorem holds for any step-function, it will also be true for any lower semi-continuous function ψ .

The main tool in the proof of Theorem 5.1 is the following new result which should be of interest in itself. For simplicity, we have also here restricted ourselves to step-functions although this result will also hold for a lower semi-continuous function p . By a solution of a differential equation or a differential inequality, we mean a function Φ such that Φ and Φ' are absolutely continuous, Φ'' is the a.e. existing derivative of Φ' and the equation or the inequality is satisfied a.e. For general properties of the kind of convexity which is considered in Theorem 5.2, we refer to Heins [2].

<u>Theorem 5.2</u>: <u>Let p</u>: $[-\infty, 0] \to [0,\infty)$ <u>be a lower semi-continuous, piecewise constant function such that the range of p consists of finitely many values</u>. <u>Assume that there exists a solution of the inequality</u>

$$\Phi''(t) - p(t)^2 \Phi(t) \geq 0 \; , \; -\infty < t \leq 0 \; , \tag{5.10}$$

<u>such that</u> $\Phi(0) = 1$, $\lim_{t\to-\infty} \Phi(t)$ <u>exists</u>.

<u>Let</u> t_0 <u>be given</u>, $t_0 < 0$. <u>If</u> $\inf_t p(t) > 0$, <u>there exists a non-negative solution</u> Φ^* <u>of the equation</u>

$$\Phi^{*''}(t) - p^*(t)^2\, \Phi^*(t) = 0 \tag{5.11}$$

such that $\Phi^*(0) = 1$, $\Phi^*(-\infty) = 0$ *and*

$$\Phi(t_o) \leq \Phi^*(t_o) . \tag{5.12}$$

Here p^* *is a measure preserving, non-decreasing rearrangement of* p *on* $[t_0, 0]$ *and* $p^*(t) = \inf_t p(t)$ *on* $(-\infty, t_0)$.

If $\inf_t p(t) = 0$, *the statement above will still be true except that the conclusion* $\Phi^*(-\infty) = 0$ *is replaced by*

$$\Phi^*(t) = \Phi^*(t_0) \ , \ t < t_0 .$$

Remark: See p. 55.

The bridge from Theorem 5.2 to Theorem 5.1 is given by the following result of F. Norstad [1]. For further applications of this technique, we refer to Essén and Shea [1].

Theorem 5A: *Let u be subharmonic in an annulus* $\{R_1 < |z| < R_2\}$. *If* λ *is given*, $0 < \lambda < 1$, *and*

$$u(-r) \leq \cos \pi\lambda \ u(r) \ , \ R_1 < r < R_2 \ ,$$

then

$$L_\lambda(r) = L_\lambda(r,u) = \int_{-\pi}^{\pi} u(re^{i\varphi}) \sin \lambda\, (\pi - |\varphi|)\, d\varphi$$

is convex with respect to the family $A\, r^\lambda + B\, r^{-\lambda}$ *(A,B are constants)*.

The case $\lambda = 1/2$ is due to L. Ahlfors (cf. Heins, Ex. 4, p. 112).

In this context, we only require a simplified version of this result. If u is a subharmonic function such that (5.1) holds in a disk $\overline{\Delta(0,R)}$, let u^* be the associated subharmonic function which is obtained by projecting the Riesz mass of u onto the negative real axis as in § 1. We know from that section that

(5.13) $$u^*(-r) \leq \cos \pi\psi(r)\, u^*(r) \;,\; 0 < r < R \;,$$

(5.14) $$u^*(re^{i\theta}) = u^*(re^{-i\theta}) \;,\; |\theta| < \pi \;,$$

(5.15) $$\theta \to u^*(re^{i\theta}) \text{ is a decreasing function on } [0,\pi].$$

In the sequel, let $L_\mu(r) = L_\mu(r,u^*)$. It is easy to check that

(5.16) $$L_\mu \text{ is continuous on } [0,R] \;.$$

(5.17) $$L'_\mu \text{ is continuous on } [0,R] \;.$$

In (5.17) it is essential that the integrand in L_μ tends to zero as $|\varphi| \to \pi$. Let $\Delta_o L_\mu = r^2 L''_\mu + rL'_\mu$.

Lemma 5.1:

(5.18) $$\Delta_o L_\mu - \mu^2 L_\mu = 2\mu\,\{u^*(r)\cos \pi\mu - u^*(-r)\} = H(r) \;,\; 0 < r < R \;.$$

Here L'_μ is absolutely continuous and L''_μ is the almost everywhere existing derivative of L'_μ . By the change of variables $r = Re^t$, we obtain

(5.19) $$\Phi''_\mu - \mu^2 \Phi_\mu = h \;,\; -\infty < t < 0 \;,$$

where $\Phi_\mu(t) = L_\mu(Re^t)$ and $h(t) = H(Re^t)$.

Our general program is as follows: we are going to use (5.13) and Lemma 5.1 to deduce a differential inequality of type (5.10). Theorem 5.2 will then give us an estimate of Φ_μ , or equivalently, L_μ . There exists an absolute constant C and a number $r_o > 0$ such that

(5.20) $$u^*(r) \leq C(L_\mu(r)\,\mu^{-1} - 2u(0)(1 - \cos \tfrac{\pi\mu}{2})\,\mu^{-2}) \;,\; r_o < r < R \;.$$

Since $u^*(r) \geq M(r)$, we obtain (5.4). We note that $u^*(0) = u(0)$ which is finite. A proof of (5.20) is given at the end of this section.

Assuming Theorem 5.2 to be true, we now give the proof of Theorem 5.1. The proof of Theorem 5.2 will be given in § 7.

Proof of Lemma 5.1: Following Norstad, we assume that u* is twice continuously differentiable and apply Green's formula to the functions u* and $\rho e^{i\theta} \to \rho^{\mu} \sin \mu(\pi - \theta)$ in the region $\{s < |z| < r\} \cap \{\text{Im } z > 0\}$ and to the functions u* and $\rho e^{i\theta} \to \rho^{\mu} \sin \mu(\pi + \theta)$ in the region $\{s < |z| < r\} \cap \{\text{Im } z < 0\}$. After addition of these two formulas we obtain that

(5.21) $$r^{\mu+1} L'_{\mu}(r) - \mu r^{\mu} L_{\mu}(r) - s^{\mu+1} L'_{\mu}(s) + \mu s^{\mu} L_{\mu}(s) =$$

$$= \int_s^r 2\mu\, t^{\mu-1}(u^*(t)\cos \pi\mu - u^*(-t))\, dt .$$

Approximating u* by a decreasing sequence of differentiable functions, we see that (5.21) holds in the general case. Hence L'_{μ} is absolutely continuous. Differentiating (5.21), we obtain (5.18), and Lemma 5.1 is proved.

By (5.18) and (5.13) we see that

(5.22) $$(\Delta_o L_{\mu} - \lambda^2 L_{\mu})(r) = (\mu^2 - \lambda^2) L_{\mu}(r) + 2\mu\{u^*(r)\cos \pi\mu - u^*(-r)\} ,$$

(5.23) $$(\Delta_o L_{\mu} - \lambda^2 L_{\mu})(r) \geq (\mu^2 - \lambda^2) L_{\mu}(r) + 2\mu u^*(r)(\cos \pi\mu - \cos \pi\psi(r))$$

The next step is provided by

Lemma 5.2 a) In the set where (5.13) holds with $\psi(r) = 0$,

(5.24) $$\Delta_o L_{\mu} \geq 0 .$$

b) In the set where (5.13) holds with $0 < \mu \leq \psi(r) < 1$,

(5.25) $$\Delta_o L_{\mu}(r) - \Psi(r)^2 L_{\mu} \geq 0 ,$$

(where Ψ is defined in Theorem 5.1).

<u>Proof</u>: a) By (5.22) with $\lambda = 0$, (5.24) is equivalent to

(5.24') $$\mu L_\mu(r) \geq 2\{u^*(-r) - u^*(r) \cos \pi\mu\}$$

If $0 \leq \mu \leq 1/2$, we use an integration by parts and (5.15) to conclude that (5.24') is true. If $0 < \mu < 1$, we first note that

(5.26) $$u^*(-r) + u^*(r) \leq u^*(-re^{i\theta}) + u^*(re^{i\theta}) , \quad |\theta| \leq \pi ,$$

$$L_\mu(r) = 2 \int_0^\pi u^*(re^{i\theta}) \sin \mu(\pi-\theta) d\theta = 2 \int_0^{\pi/2} \{u^*(-re^{i\theta}) + u^*(re^{i\theta})\} \sin \mu\theta \, d\theta$$

$$+ 2 \int_0^{\pi/2} u^*(re^{i\theta}) \{\sin \mu (\pi - \theta) - \sin \mu\theta\} d\theta = I_1 + I_2 .$$

Using (5.26), we see that

$$\mu I_1 \geq 2(u^*(-r) + u^*(r))(1 - \cos (\pi\mu/2)) .$$

Applying an integration by parts, it follows from (5.15) that

$$\mu I_2 \geq u^*(ir) \, 4 \cos (\pi\mu/2) - 2u^*(r)(1 + \cos \pi\mu) .$$

Once more using (5.26), this time with $\theta = \pi/2$, we see that

$$\mu(I_1 + I_2) \geq 2(u^*(-r) - \cos \pi\mu \, u^*(r)) ,$$

and (5.24) is proved.

b) Using (5.15), we see that

$$\mu L_\mu(r) \leq u^*(r)(1 - \cos \pi\mu) .$$

Plugging this estimate into (5.23), with $\lambda = \Psi(r)$, we obtain (5.25). The proof of Lemma 5.2 is complete.

Proof of Theorem 5.1: It follows from Lemma 5.2 that (5.24) and (5.25) hold in the set where $\psi(r) = 0$ and $\mu \leq \psi(r) < 1$, respectively. Changing variables as in Lemma 5.1, we obtain

$$\Phi''_\mu(t) - p(t)^2 \Phi_\mu(t) \geq 0 , -\infty < t < 0 \tag{5.27}$$

where $p(t) = \Psi(Re^t)$, $-\infty < t < 0$, fulfills the assumptions of Theorem 5.2. We note that $\Phi_\mu(0) = 2M(R)(1 - \cos \pi\mu)\mu^{-1}$ and that $\lim_{t\to-\infty} \Phi(t) \leq 0$. If $t_0 = \log (r/R) < 0$ is given, we use Theorem 5.2 and (5.12) gives us an estimate of $\Phi_\mu(t_0)$. In Lemma 5.3 below, there is an **estimate of the** solution of the re-arranged equation (5.11). The final conclusion in Theorem 5.1, (5.4), is now an immediate consequence of (5.20) and Lemma 5.3.

Lemma 5.3: Let $0 \leq \sigma_1 < \sigma_2 < \dots \sigma_q$ and $\{\Delta_k\}_1^q$ be given, $\Delta_k > 0$, $k = 1,2,\dots q$. Let

$$p^*(t) = \sigma_k , \sum_1^{k-1} \Delta_\nu < t \leq \sum_1^k \Delta_\nu, \; k = 2,3,\dots$$

$$p^*(t) = \sigma_1 , 0 \leq t \leq \Delta_1 .$$

Let Φ be a solution of equation (5.11) in $(0, \sum_1^q \Delta_k)$ such that $\Phi(0) = 1$, $\Phi'(0) = \sigma_1$. Then

$$\Phi(\Delta_1 + \Delta_2) \geq \frac{1}{2} (1 + (\sigma_1/\sigma_2)) \exp \{\sigma_1 \Delta_1 + \sigma_2 \Delta_2\} \tag{5.28}$$

$$\Phi(\sum_1^q \Delta_k) \geq \frac{1}{4} ((\sigma_1 + \sigma_2)/\sigma_q) \exp \{\sum_1^q \sigma_k \Delta_k\} . \tag{5.29}$$

Proof: We note that Φ and Φ' are continuous. By computing Φ in $(0,\Delta_1 + \Delta_2)$, it is easy to see that (5.28) holds. To prove (5.29), let $\Phi (\sum_1^k \Delta_j) = a_k$, $\Phi' (\sum_1^k \Delta_j) = b_k$. Then

(5.30) $$a_{k+1}\,\sigma_{k+1} + b_{k+1} = (a_k\,\sigma_{k+1} + b_k)\exp\{\sigma_{k+1}\,\Delta_{k+1}\} >$$

$$> (a_k\,\sigma_k + b_k)\exp\{\sigma_{k+1}\,\Delta_{k+1}\}\ ,\ k = 1,2,\dots q-1\ .$$

(5.31) $$a_k\,\sigma_k > b_k\ ,\ k = 2,3,\dots q\ .$$

By multiplying together the inequalities defined in (5.30) and using (5.28) and (5.31), we obtain (5.29).

<u>Proof of Corollary 1</u>: We define

$$\psi(r) = \begin{cases} \lambda\ , & r \in E\ , \\ 0\ , & r \in (0,\infty)\setminus E\ . \end{cases}$$

In this case, the smallest positive number in the range of ψ is λ. It follows that $\Psi = \psi$. The corollary is now an immediate consequence of (5.1) and (5.4).

<u>Proof of Corollary 2</u>: We introduce

$$E_2 = F_\rho = \{r\colon m(r) \le \cos\pi\rho\ M(r)\}\ ,$$

$$E_1 = F_{\rho_0}\setminus F_\rho = \{r\colon \cos\pi\rho\ M(r) < m(r) \le \cos\pi\rho_0\ M(r)\}\ ,$$

$$E_0 = (0,\infty)\setminus E_1 = \{r\colon \cos\pi\rho_0\ M(r) < m(r) \le M(r)\}\ .$$

Hence (5.1) will be true with

$$\psi(r) = \begin{cases} \rho\ , & r \in E_2 \\ \rho_0\ , & r \in E_1 \\ 0\ , & r \in E_o \end{cases}$$

The smallest positive number in the range of ψ is ρ_0 .

There exists a positive constant $C(\rho_o)$ such that

$$\Psi(r) = \begin{cases} \rho_o \{(1-\cos \pi\rho)/(1-\cos \pi\rho_o)\}^{1/2} \geq \rho_o + C(\rho_o)(\rho-\sigma_o), & r \in E_2 \\ \rho_o , & r \in E_1 \\ 0 , & r \in E_o \end{cases}$$

If $\underline{\Lambda}\, F_{\rho_o} = 1$, $\overline{\Lambda}\, E_o = 0$. Let $\overline{\Lambda}\, F_\rho = \alpha$. Taking logarithms in (5.4) dividing by log R and letting $R \to \infty$ through a sequence $\{R_\nu\}_1^\infty$ such that $m_\ell\, F_\rho(r,R_\nu)/\log R_\nu \to \alpha$, $\nu \to \infty$, we see that

$$\rho_o + C(\rho_o)(\rho - \rho_o)\,\alpha \leq \rho_o \,.$$

It follows that $\alpha = 0$, e.g., we have proved (5.5).

To prove the analogue where $\overline{\Lambda}$ and $\underline{\Lambda}$ are interchanged, assume that $\overline{\Lambda}\, F_{\rho_o} = 1$. It follows that $\underline{\Lambda}\, E_o = 0$.

Let $\underline{\Lambda}\, F_\rho = \beta$. Arguing as above, except that the sequence $\{R_\nu\}$ is chosen such that $m_\ell\, E_o(r,R_\nu)/\log R_\nu \to 0$, $\nu \to \infty$, it follows that $\beta = 0$, e.g., (5.6) is true.

To prove (5.8), let if $\rho > \rho_o$ and $\varepsilon > 0$ are given $\lambda = \rho_o - \varepsilon(\rho - \rho_o)$. Let E_2 be as in the previous argument, let $E_1 = F_\lambda \setminus F_\rho$ and $E_o = (0,\infty)\setminus E_1$.

Hence (5.1) will be true with

$$\psi(r) = \begin{cases} \rho , & r \in E_2 , \\ \lambda , & r \in E_1 , \\ 0 , & r \in E_o . \end{cases}$$

$$\Psi(r) \quad \begin{cases} \geq \lambda + C(\lambda)(\rho - \lambda) , & r \in E_2 , \\ = \lambda , & r \in E_1 , \\ = 0 , & r \in E_o . \end{cases}$$

From (5.7), we see that $\bar{\Lambda}\, E_o = 0$. Using (5.4) in the same way as above, it follows that

$$C(\lambda)(\rho - \lambda)\, \bar{\Lambda}\, E_2 + \lambda \leq \rho_o ,$$

e.g., $\bar{\Lambda}\, E_2 = \bar{\Lambda}\, F_\rho \leq \varepsilon\, \{C(\lambda)(1 + \varepsilon)\}^{-1}$. Since $\varepsilon > 0$ is arbitrary, $\bar{\Lambda}\, F_\rho = 0$, $\rho_o < \rho < 1$. It is now easily seen that (5.8) is true.

The proof of Corollary 3 is similar and is omitted.

<u>Proof of (5.20)</u>:

$$L_\mu(r) = \int_0^{\pi/2} \{u^*(re^{i\theta}) + u^*(-re^{i\theta})\} \sin\mu\theta\, d\theta + \int_0^{\pi/2} u^*(re^{i\theta})(\sin\mu(\pi-\theta) - \sin\mu\theta)\, d\theta =$$

$$= I_1 + I_2 .$$

Since $\theta \to u^*(re^{i\theta}) + u^*(-re^{i\theta})$ is increasing on $[0,\pi/2]$ and u^* is subharmonic,

$$I_1 \{ \int_0^{\pi/2} \sin \mu\theta\, d\theta \}^{-1} \geq \frac{2}{\pi} \int_0^{\pi/2} (u^*(re^{i\theta}) + u^*(-re^{i\theta}))\, d\theta \geq u^*(0)/2 = u(0)/2 .$$

We note that $\sin \mu(\pi - \theta) - \sin \mu\theta \geq 0$, $0 \leq \theta \leq \pi/2$ and that there is a number $r_o > 0$ such that $u^*(re^{i\theta}) > 0$, $0 \leq \theta \leq \pi/2$, $r \geq r_o$. Applying Harnack's inequality, we obtain a lower bound for $I_2/u^*(r)$. Combining these estimates, we obtain (5.20).

<u>Remark</u>: Theorem 5.2 can also be used in the study of a differential inequality of Carleman (cf. Heins p. 121; in Haliste [1], Lemma 6.3, the same result is given in a different form). We would like to indicate how the methods of these notes can be used in connection with the problem discussed in § 6 in Haliste [1]: to find estimates for certain harmonic measures. We use the notation of Haliste. Let D be a domain in $\underline{R}^n$ satisfying condition A in Haliste p. 13 except that we do not have to assume that D is simply connected. We first apply the generalization of a result of Baernstein given by Borell at the end of these notes and replace D by its Schwarz symmetrization D* with respect to the x_1 axis: the new Carleman mean is a majorant of the original one, and D* is a simply connected domain. If φ is the Carleman mean defined as in (6.1) in Haliste [1]

and $\psi = \sqrt{\varphi}$, we see from Lemma 6.4 in Haliste that

$$\psi''(x) - \lambda^*(x)\,\psi(x) \geq 0$$

where $\lambda^*(x)$ is the principal eigenvalue of the sphere $D^* \cap \{x_1 = x\}$. The connection between the radius $\rho(x)$ of this sphere and $\lambda^*(x)$ is well-known. As an example, we mention that when the dimension $n = 2$, $\lambda^*(x) = \pi^2/(2\rho(x))^2$. Theorem 5.2 (in the form given in (7.14)) might be used to deduce generalizations of Theorems 6.2 and 6.3 in Haliste [1]. We note that the restrictions (a), (b) and (c) in Haliste's Theorem 6.3 are no longer needed. It follows from condition A in Haliste [1] that $\rho(x)$ is bounded when x is large. If we assume only that $\rho(x)$ is locally bounded, we can still obtain an estimate using Theorem 5.2 and Lemma 5.3.

Christer Borell has remarked that these methods apply also to cap symmetrizations. To obtain this result, he uses Theorem 4.29 in a paper of Sarvas which can be found in the bibliography of Borell's paper in these notes.

6. A counterexample

We here give examples of functions subharmonic in $\underline{C}$ for which Kjellberg's conjecture is not fulfilled.

Let λ, η and A be given positive numbers such that $1/2 < \lambda < 1$ and $\eta A < \lambda$. Consider

$$u(z) = \operatorname{Re} \{\int_0^\infty z(z+t)^{-1} \nu(t)\, dt/t\}, \tag{6.1}$$

where

$$\nu(t) = \exp \{\int_1^t \{\lambda + \eta A \cos(A \log s)\} ds/s\} = t^\lambda \exp\{\eta \sin(A \log t)\}.$$

It is clear that ν is nondecreasing on $(0,\infty)$ and that u is subharmonic in $\underline{C}$. We define, if $r > 0$,

$$\cos \pi\varkappa(r) = u(-r)/u(r) = \tag{6.2}$$

$$\{\int_0^\infty s^{\lambda-1}(s-1)^{-1}\exp\{\eta \sin(A \log(rs))\}ds\}\{\int_0^\infty s^{\lambda-1}(s+1)^{-1}\exp\{\eta \sin(A \log(rs))\}ds\}^{-1},$$

where the integral in the numerator should be interpreted as a principal value and $0 < \varkappa(r) < 1$. Expanding the exponential terms in power series in η, we find that

$$(\log R)^{-1} \int_1^R \varkappa(r)dr/r = \lambda - \eta^2 C(\lambda,A) + \underline{O}(1)(\log R)^{-1} + \eta^3),\ R \to \infty, \tag{6.3}$$

where

$$C(\lambda,A) = (4\pi)^{-1}\cot \pi\lambda(\cosh \pi A - 1)^2 (\sin^2\pi A + (\sinh \pi A)^2)^{-1} < 0.$$

It is easy to see that

$$M(r,u) = u(r) \leq \int_0^\infty r(r+t)^{-1} t^{\lambda-1} e^{\eta} dt = \pi r^\lambda e^{\eta}/\sin \pi\lambda ,$$

$$M(r,u) = u(r) \geq \int_0^\infty r(r+t)^{-1} t^{\lambda-1} e^{-\eta} dt = \pi r^\lambda e^{-\eta}/\sin \pi\lambda .$$

Hence, if $r < R$,

(6.4) $$e^{-2\eta} (r/R)^\lambda \leq M(r)/M(R) \leq e^{2\eta} (r/R)^\lambda .$$

If the conjecture of Kjellberg mentioned in § 5 is true,

(6.5) $$M(r) \leq \text{Const. } M(R) \exp \{- \int_r^R \varkappa(t)\, dt/t\} , \quad r < R .$$

Taking logarithms in (6.5), dividing by log R and letting $R \to \infty$, we see from (6.3) - (6.5) that

(6.6) $$0 \leq \lambda - \lambda + \eta^2 C(\lambda, A) + C_1 (\lambda, A) \eta^3 .$$

Since $C(\lambda, A) < 0$, we can choose η so small so that (6.6) is not true. Hence (6.5) is also false and we have a counterexample for functions subharmonic in $\underline{C}$. The order of magnitude of the term $\eta^2 C(\lambda, A)$ in (10.6) which causes this effect is, if A is large,

(6.7) $$\eta^2 (4\pi)^{-1} \cot \pi\lambda \approx \text{Const. (the deviation from } \lambda)^2 .$$

The estimate which can be found in Essén [3] (cf. the discussion in §5) differs from the conjecture of Kjellberg by a term which is of the same order of magnitude as (6.7).

We can obtain a more complicated increasing function than the one defined in (6.2) by taking

$$\nu(r) = r^\lambda \exp \{\eta \sum_1^q B_k \sin (A_k \log r)\} ,$$

where $\{A_k\}$, $\{B_k\}$ and η are chosen in such a way that

$$0 < \eta \sum_1^q |A_k B_k| < \lambda .$$

Then (6.3) is replaced by

$$(6.3') \qquad (\log R)^{-1} \int_1^R \varkappa(r)\ dr/r = \lambda - C_1(\lambda, \{A_k\})\eta^2 \sum_1^q B_k^2 + $$

$$+ \underline{O}(1)((\log R)^{-1} + \eta^3) \ , \ R \to \infty \ , \text{ where } C_1(\lambda, \{A_k\}) < 0 .$$

By choosing the trigonometric polynomial in the right way, we obtain a function $\varkappa$ which will be an approximation of a periodic, piecewise constant function.

If we want our counterexample to be the logarithm of the modulus of an entire function, we define first $n(r) = [\nu(r)]$ and secondly

$$v(z) = \text{Re} \ \{ \int_0^\infty z(z + t)^{-1} \ n(t) \ dt/t\} .$$

Using the technique of Hayman [2, p. 154 - 155], it is easy to see that v is so close to u that our argument works also for the function $\varkappa(\circ, v)$ defined by

$$\cos \pi \varkappa (r, v) = \max \{v(-r)/v(r) \ , -1\} .$$

7. Proof of Theorem 5.2

Consider the differential equation

$$z''(t) - p(t)^2 z(t) = 0 , \tag{7.1}$$

where p is a piecewise constant, nonnegative function and a solution is defined as in Theorem 5.2. From a theorem of Kneser (cf. Kneser [1], a similar result is given in Bebernes and Jackson [1]) we conclude that there exists a unique nonnegative solution of (7.1) on $(-\infty,0]$ such that $\lim_{t\to-\infty} z(t)$ exists. From Lemma 7.1 below, we see that z majorizes Φ in (5.10). From now on, we consider z. Let t_0, t_1 and t_2 be given, $t_0 < t_1 < t_2$, and define

$$\begin{cases} p_1(t) = p(t) , & t < t_0 , \ t > t_2 , \\ p_1(t) = p(t + t_1 - t_0) , & t_0 < t < t_0 + t_2 - t_1 , \\ p_1(t) = p(t - t_2 + t_1) , & t_0 + t_2 - t_1 < t < t_2 . \end{cases}$$

Let us say that we have switched the intervals (t_0, t_1) and (t_1, t_2) in the differential equation if instead of z, we consider z_1 which satisfies

$$z_1'' - p_1(t)^2 z_1 = 0 , \tag{7.2}$$

$z_1(t_0) = z(t_0)$, $z_1'(t_0) = z'(t_0)$. In particular, $z_1(t) = z(t)$, $t \le t_0$.

We start with a number of lemmas. In the sequel, p is always a nonnegative step function with a finite number of jumps on each compact set. Let $\sigma = \inf_t p(t)$. Our method of proof works well in the case $\sigma = 0$. For simplicity, we only discuss the case $\sigma > 0$ and leave the simple modifications which are necessary when $\sigma = 0$ to the reader.

Lemma 7.1: If T is given, $T < 0$, and if z satisfies

(7.3) $$z'' - p(t)^2 z \geq 0 , \; z(0) = 0 , \; z(T) = 0 ,$$

then $z(t) \leq 0$, $T \leq t \leq 0$. If $T = -\infty$, this statement is true if we define $z(-\infty) = \lim_{t \to -\infty} z(t)$.

We omit the proof.

Lemma 7.2: Let σ_1, σ_2 be given $\sigma_1 > \sigma_2 \geq 0$. Let Δ_1 and Δ_2 be positive numbers and consider the solutions of the following differential equations.

$$\begin{cases} z_1'' - \sigma_1^2 z_1 = 0 , & 0 < t < \Delta_1 , \\ z_1'' - \sigma_2^2 z_2 = 0 , & \Delta_1 < t < \Delta_1 + \Delta_2 , \end{cases}$$

$$\begin{cases} z_2'' - \sigma_2^2 z_2 = 0 , & 0 < t < \Delta_2 , \\ z_2'' - \sigma_1^2 z_2 = 0 , & \Delta_2 < t < \Delta_1 + \Delta_2 , \end{cases}$$

$$z_1(0) = z_2(0) = a > 0 , \quad z_1'(0) = z_2'(0) = b > 0 .$$

Then

$$z_2(\Delta_1 + \Delta_2) < z_1(\Delta_1 + \Delta_2) ,$$

$$z_2'(\Delta_1 + \Delta_2) > z_1'(\Delta_1 + \Delta_2) ,$$

$$z_1(\Delta_1 + \Delta_2) - z_2(\Delta_1 + \Delta_2) = (a/b)(z_2'(\Delta_1 + \Delta_2) - z_1'(\Delta_1 + \Delta_2)) .$$

Proof: By explicit computation.

Lemma 7.3: Let $\sigma > 0$ be given and assume that $p(t) \geq \sigma > 0$, $t > 0$. Let z_1, z_2 be solutions of (7.1) such that $z_1(0) = A_1$, $z_1'(0) = B_1$, $z_2(0) = A_2 < A_1$, $z_2'(0) = B_2 > B_1$, where $\sigma(A_1 - A_2) \geq B_2 - B_1$. Then the graphs of the two functions z_1 and z_2 never intersect when $t > 0$.

Proof: Consider $w = z_1 - z_2$ which satisfies (7.1). We see that $w(0) = A_1 - A_2 = A > 0$, $w'(0) = B_1 - B_2 = -B < 0$, where $\sigma A \geq B$. Assume that the graph of w cuts the positive t-axis at $t_o > 0$ and that $w(t) > 0$, $0 < t < t_o$. Let v be the solution of $v'' - \sigma^2 v = 0$, $v(0) = A$, $v'(0) = -B$. Since $(w-v)'' - \sigma^2(w-v) \geq 0$, $0 < t < t_o$, it follows from Lemma 7.1 that the graph of v cuts the positive t-axis at t_1, $0 < t_1 < t_o$. But

$$v(t) = A \cosh \sigma t - \sigma^{-1} B \sinh \sigma t ,$$

and $v(t_1) = 0$ is equivalent to $\tanh \sigma t_1 = A\sigma/B \geq 1$. But $\tanh \sigma t < 1$, $t > 0$, we have a contradiction and there can be no point $t_o > 0$ such that $w(t_o) = 0$. Lemma 7.3 is proved.

We want to investigate what happens to a solution of (7.1) when an interval where p is minimal is shifted to the left.

Lemma 7.4: Let σ, Δ, Δ_1 be given positive numbers. Assume that

$$\begin{cases} p(t) = \sigma , & t < 0 , \quad \Delta < t < \Delta + \Delta_1 , \\ p(t) \geq \sigma , & -\infty < t < \infty . \end{cases}$$

Let p_1 be the function obtained by switching the intervals $(0,\Delta)$ and $(\Delta, \Delta + \Delta_1)$. Let z_1 be a solution of (7.1) such that $z_1(t) = e^{\sigma t}$, $t < 0$. Let z_2 be a solution of (7.2) such that $z_2(t) = e^{\sigma t}$, $t < 0$. Then $z_1(t) > z_2(t)$, $t > \Delta + \Delta_1$.

Proof: By computing the derivative of $g(t) = e^{\sigma t}\{\sigma z_1(t) - z_1'(t)\}$ and using (7.1), we see that g is a nonincreasing function. In particular,

$$\sigma z_1(\Delta) - z_1'(\Delta) \leq e^{-\sigma\Delta} (\sigma z_1(0) - z_1'(0) = 0 .$$

Solving the differential equations, we see that

$$z_1(\Delta + \Delta_1) = z_1(\Delta) \cosh \sigma\Delta_1 + \sigma^{-1} z_1'(\Delta) \sinh \sigma\Delta_1 ,$$

$$z_1'(\Delta + \Delta_1 = \sigma z_1(\Delta) \sinh \sigma\Delta_1 + z_1'(\Delta) \cosh \sigma\Delta_1 ,$$

and that

$$z_2(\Delta + \Delta_1) = z_1(\Delta) e^{\sigma\Delta_1} , z_2'(\Delta + \Delta_1) = z_1'(\Delta) e^{\sigma\Delta_1} .$$

Hence

$$z_1(\Delta + \Delta_1) - z_2(\Delta + \Delta_1) = \sinh \sigma\Delta_1(\sigma^{-1} z_1'(\Delta) - z_1(\Delta)) \geq 0 ,$$

$$z_2'(\Delta + \Delta_1) - z_1'(\Delta + \Delta_1) = \sinh \sigma\Delta_1(z_1'(\Delta) - \sigma z_1(\Delta)) \geq 0 .$$

Lemma 7.4 is now an immediate consequence of Lemma 7.3.

We can now take the first step in the proof of Theorem 5.2. We want an estimate at $t_o < 0$ of a solution z of (7.1) such that $z(0) = 1$, $\lim_{t \to -\infty} z(t) \leq 0$. If $z(t_o) \leq 0$, the estimate of Theorem 5.2 is trivially true. From now on, we assume that $z(t_o) > 0$.

We shall, step by step, replace z by solutions $\{z_k\}$ of differential equations such that $z_k(t_o) = z(t_o)$, $0 < z_k(t) \leq z(t)$ in some interval $(t_k,0)$, $k = 1,2,\ldots$ Clearly

$$z(t_o) \leq z_k(t_o) / z_k(0)$$

which is the conclusion of Theorem 5.2.

I. It follows from Lemma 7.1 that $z(t) \leq z(t_o)e^{\sigma t}$, $t \leq t_o$, where $\sigma = \inf_{t<0} p(t) > 0$. In particular, we see that $z'(t_o) \geq \sigma z(t_o)$.
Let z_1 be a solution of (7.1) in $(t_o,0)$ such that $z_1(t_o) = z(t_o)$, $z_1'(t_o) = \sigma z(t_o)$. It is obvious that $0 < z_1(t) \leq z(t)$, $t_o \leq t \leq 0$.

II. Assume that

$$\begin{cases} p(t) > \sigma , & t_o < t < t_o + \Delta , \\ p(t) = \sigma , & t_o + \Delta < t < t_o + \Delta + \Delta_1 . \end{cases}$$

Using Lemma 7.4, we switch these two intervals and obtain a function z_2 such that $0 < z_2(t) \leq z_1(t)$, $t_o + \Delta + \Delta_1 < t < 0$. Continuing in the same way, we push all the σ-intervals of p as far to the left in the interval $(t_o,0)$ as we can. We obtain a rearrangement P of p such that the solution Z of the new equation which has the same initial values at t_o as z_1 is such that

$$0 < Z(t) \leq z_1(t) \leq z(t) \ , \ t_o + \Delta(\sigma) < t < 0 \ .$$

Here $\Delta(\sigma)$ is the total length of the σ-intervals in $(t_o,0)$ of the function p. If the range of p is $\{\sigma,\rho\}$, where $0 < \sigma < \rho$, we are done, because

$$P(t) = \begin{cases} \sigma , & t_o < t < t_o + \Delta(\sigma) , \\ \rho , & t_o + \Delta(\sigma) < t < 0 , \end{cases}$$

is the non-decreasing rearrangement of p in $(t_o,0)$.

III. If the range of p contains more than two numbers $\sigma_0, \sigma_1, \sigma_2, \ldots$, where $0 < \sigma_0 < \sigma_1 < \sigma_2 < \ldots$, we have to do more of the same. We first use the previous argument to push all the σ_0-intervals as far to the left in $(t_o,0)$ as possible. Let us put $t_1 = t_o + \Delta(\sigma_0)$. We now want to push the σ_1-intervals as far to the left in the interval $(t_1,0)$ as possible. We need the following lemma.

Lemma 7.5: Assume that $\Delta, \Delta_1, \Delta_2$ are given positive numbers. Put $\tilde{\Delta} = \Delta + \Delta_1 + \Delta_2$. Let

$$
\begin{cases}
p(t) = \begin{cases} \sigma_o , & t < 0 , \\ \sigma_1 , & 0 < t < \Delta_1 , \ \Delta + \Delta_1 < t < \tilde{\Delta} \end{cases} \\
p(t) \geq \sigma_1 , \quad t \geq 0 ,
\end{cases}
$$

and let p_1 be the function obtained by switching the intervals $(\Delta_1 , \Delta_1 + \Delta)$ and $(\Delta_1 + \Delta , \tilde{\Delta})$. Let z_1 and z_2 be solutions of (7.1) and (7.2), respectively which both are $\exp(\sigma_o t)$, $t < 0$ (for simplicity, we take $t_o = 0$). If

(7.4) $$z_1(\tilde{\Delta}) \geq z_2(\tilde{\Delta}) ,$$

(7.5) $$\sigma_1\{z_1(\tilde{\Delta}) - z_2(\tilde{\Delta})\} \geq z_2'(\tilde{\Delta}) - z_1'(\tilde{\Delta}) ,$$

then $z_2(0) = z_1(0)$ and $z_2(t) \leq z_1(t)$, $\tilde{\Delta} \leq t$. (If (7.4) holds but (7.5) is false, then the last conclusion need not be true.)

Lemma 7.5 is an immediate consequence of Lemma 7.3. (It is easy to find simple counterexamples if (7.5) is not true.) If (7.4) and (7.5) hold, we can go on pushing the σ_1-intervals to the left and obtain a decreasing sequence of solutions of the associated differential equations. The remaining problem is to check when (7.4) and (7.5) are true. The main tool is the following

Lemma 7.6: Let p and p_1 be as in Lemma 7.5 and let v_1 and v_2 be solutions of (7.1) in $(\Delta_1 , \Delta_1 + \Delta) = (a,b)$ such that

$$v_1(a) = 1 , \ v_1'(a) = 0 , \ v_2(a) = 0 , \ v_2'(a) = 1 ,$$

$$v_1(b) = A_1 , \ v_1'(b) = B_1 , \ v_2(b) = A_2 , \ v_2'(b) = B_2 .$$

Then

(7.6) $$B_1 - \sigma_1^2 A_2 \geq \sigma_1 |B_2 - A_1| .$$

Proof: If v is a solution of (7.1) in (a,b), we see by differentiating and using (7.1) that in (a,b)

$$(7.7) \qquad e^{\sigma_1 t}(\sigma_1 v - v') \text{ is non-increasing },$$

$$(7.8) \qquad e^{-\sigma_1 t}(\sigma_1 v + v') \text{ is non-decreasing }.$$

If we choose $v = v_1 + \sigma_1 v_2$ in (7.7) we see that

$$\sigma_1(v_1+\sigma_1 v_2)(b) - (v_1+\sigma_1 v_2)'(b) \leq \exp\{\sigma_1(a-b)\}(\sigma_1 v - v')(a) = 0 .$$

If we choose $v = v_1 - \sigma_1 v_2$ in (7.8), we see that

$$\sigma_1(v_1-\sigma_1 v_2)(b) + (v_1-\sigma_1 v_2)'(b) \geq \exp\{\sigma_1(b-a)\}(\sigma_1 v + v')(a) = 0 .$$

Combining these two inequalities, we obtain (7.6). Lemma 7.6 is proved.

To prove that (7.4) and (7.5) are true, we see (after some computation) that if z_1 and z_2 are defined as in Lemma 7.5, then

$$(7.9) \quad \sigma_1(z_1(\tilde{\Delta}) - z_2(\tilde{\Delta})) = \sinh \sigma_1\Delta_2 \{C_1(B_1 - A_2\sigma_1^2) + D_1(B_2 - A_1)\} .$$

$$(7.10) \quad \sigma_1(z_2'(\tilde{\Delta}) - z_1'(\tilde{\Delta})) = \sinh \sigma_1\Delta_2 \{\sigma_1^2 C_1(B_2 - A_1) + D_1(B_1 - A_2\sigma_1^2)\} ,$$

and that

$$(7.11) \qquad \sigma_1(z_1(\tilde{\Delta}) - z_2(\tilde{\Delta})) - (z_2'(\tilde{\Delta}) - z_1'(\tilde{\Delta})) =$$

$$= \sinh \sigma_1\Delta_2 (C_1 - \sigma_1^{-1} D_1)(B_1 - A_2\sigma_1^2 - \sigma_1(B_2 - A_1)) ,$$

where $C_1 = \cosh \sigma_1\Delta_1 + \sigma_1^{-1}\sigma_0 \sinh \sigma_1\Delta_1$, $D_1 = \sigma_1 \sinh \sigma_1\Delta_1 + \sigma_0 \cosh \sigma_1\Delta_1$

We note that

(7.12) $$C_1\sigma_1 - D_1 = (\sigma_1 - \sigma_0)\exp(-\sigma_1\Delta_1) > 0 .$$

Using (7.6), (7.9), (7.11) and (7.12), we see that (7.4) and (7.5) are true. This means that we can push all the σ_1-intervals as far to the left as possible in the interval $(t_0 + \Delta(\sigma_0), 0)$. Let $t_2 = t_0 + \Delta(\sigma_0) + \Delta(\sigma_1)$. We note that after this rearrangement, $\inf_{t>t_2} p(t) = \sigma_2$. Repeating this argument, we push all the σ_2-intervals as far to the left as possible in the interval $(t_2, 0)$. Continuing in this way, we obtain the measure-preserving, non-decreasing rearrangement p^* of p and the proof of Theorem 5.2 is complete.

If we furthermore assume that $\sigma = \inf_t p(t) > 0$, there is a shorter proof of an estimate which can be used instead of Theorem 5.2. We note that we also know what happens when $\sigma = 0$.

Alternative proof when $\sigma > 0$.

Let z be a solution of (7.1) such that $z(0) = 1$ and $z(-\infty) = 0$. In this context, let us assume that p is a non-negative, locally integrable function on $(-\infty, 0)$. We want an estimate of $z(t_0) = a > 0$ where $t_0 < 0$ is given. Let $z'(t_0) = b > 0$. Putting $z = e^u$, we obtain

$$u'' + (u')^2 = p^2 \quad , \quad u(t_0) = \log a = \alpha \quad , \quad u'(t_0) = b/a = \beta .$$

Using an argument of Carleman (cf. Carleman [1] or Heins [2], 65 - 66) we see that since $u' > 0$,

$$u' + \frac{u''}{2u'} \geq p \quad , \quad t \geq t_0 .$$

After two integrations, we obtain

$$\exp(2u(t)) - e^{2\alpha} \geq 2\beta\, e^{2\alpha} \int_{t_0}^{t} \exp\{2P(s)\}\, ds \quad , \quad t \geq t_0 ,$$

where $P(t) = \int_{t_o}^{t} p(s)\,ds$.

Going back to our original notation, we see that

$$z(t) \geq \sqrt{2ab}\,\exp\{P(t-1)\}\ ,\ t > t_o + 1\ . \tag{7.13}$$

(Since $p \geq 0$, it is clear that $\int_{t_o}^{t} \exp\{2P(s)\}ds \geq \exp\{2P(t-1)\}$, $t > t_o + 1$) .

It remains to estimate $b/a = z'(t_o)/z(t_o)$. Since $p(t) \geq \sigma > 0$, it is clear from Lemma 7.1 that the function v defined by

$$v'' - \sigma^2 v = 0\ ,\ v(-\infty) = 0\ ,\ v(t_o) = z(t_o)\ ,$$

is a majorant of z on $(-\infty, t_o)$. Hence $z'(t_o)/z(t_o) \geq v'(t_o)/v(t_o) = \sigma$. From (7.13) we see that

$$z(0)/z(t_o) \geq \sqrt{2\sigma}\,\exp\{P(-1)\}\ ,$$

or equivalently,

$$z(t_o) \leq \exp\{-\int_{t_o}^{-1} p(s)\}/\sqrt{2\sigma}\ . \tag{7.14}$$

Thus we have an estimate of $z(t_o)$ which can be used instead of Theorem 5.2 when $\sigma > 0$. The case $\sigma = 0$ is not covered by this argument.

8. Two consequences of a theorem of A. Baernstein

This section contains joint work of M. Essén and J. Lewis.

We wish to extend Theorem 5.1 and its corollaries to the situation considered by A. Baernstein [2] which was briefly discussed at the end of § 2. If u is subharmonic in $\underline{C}$, let

$$m_\beta(r,\alpha,u) = \inf_{|\varphi|<\beta} u(re^{i(\varphi+\alpha)}) ,$$

$$\sup_\alpha m_\beta(r,\alpha,u) = m_\beta(r,u) = m_\beta(r) .$$

As in the discussion of Theorem 5.1, we assume for simplicity that u is harmonic in $\Delta(0,1)$.

Theorem 8.1: Let u be as above and let $\psi:[0,\infty) \to [0,\infty)$ be a lower semi-continuous step-function taking finitely many values. Let β be given such that

$$0 < \beta \leq \pi \quad , \quad 0 \leq \beta\psi(r) < \pi .$$

Let the smallest positive value in the range of ψ be μ. We define $\Psi_\beta(r) = \mu\,\{(1 - \cos\beta\psi(r)/(1 - \cos\beta\mu)\}^{1/2}$. If

$$(8.1) \qquad m_\beta(r) \leq \cos\beta\psi(r)\, M(r) \ , \ r > 0 ,$$

then (with constants as in (5.4))

$$(8.2) \qquad M(r) \leq \text{Const.}(|u(0)| + M(R)\exp\{-\int_r^R \Psi_\beta(t)dt/t\}) \ , \ r_0 < r < R .$$

If we have Theorem 8.1, it is easy to state and to prove analogoues of Corollaries 1-3 of Theorem 5.1. We state the analogue of the second part of Corollary 2, since it answers in part a question raised by Baernstein ([2], § 1): can certain results of Drasin and Shea [1] be extended to the more general situation considered here?

Corollary 2: *Let* ρ_0 *and* $\overline{\lambda}$ *be defined as in § 5. If* $\beta > 0$ *is given,* $0 < \beta\rho_0 < \pi$, *and*

$$\limsup_{r\to\infty} m_\beta(r)/M(r) \leq \cos \beta\rho_0 ,$$

then

$$\overline{\lambda} \{r: m_\beta(r) \leq \cos \beta\rho \; M(r)\} = 0 , \quad \rho_0 < \rho < \pi/\beta ,$$

and there exists a set $G(\rho_0)$, $\overline{\lambda}G(\rho_0) = 0$, *such that*

$$m_\beta(r)/M(r) \to \cos \beta\rho_0 , \; r \to \infty , \; r \notin G(\rho_0) .$$

Starting from the functions introduced by Baernstein, we construct a function subharmonic in a disk to which we can apply Theorem 5.1. We need the following lemma. We remind the reader that

$f_\theta(re^{i\theta}) = \frac{\partial}{\partial\theta} f(re^{i\theta})$.

Lemma 8.1: *Let* $r \to V(r)$, $0 < r \leq 2A$, *be a bounded nondecreasing convex function of* $\log r$ *such that* $V(A)$ *is positive and let* B *be a positive constant. Let* G *be a bounded harmonic function in* $\Delta(0,A) \cap \{\mathrm{Im}\, z > 0\}$ *with the following boundary values:*

$$G(r) = 0 \;, \quad G(-r) = V(r) \;, \quad 0 < r < A .$$

$$G(Ae^{i\theta}) = \begin{cases} 2\pi B , & 0 < \theta < \pi/2 , \\ 4\pi B , & \pi/2 < \theta < \pi . \end{cases}$$

Using the reflection principle, we extend G to $D = \{z: 0 < |z| < A , |\arg z| < \pi\}$. Then $U = G_\theta$ is the restriction to D of a function subharmonic in $\Delta(0,A)$ and for each $\varkappa$, $0 < \varkappa < 1$, there exists a constant only depending on $\varkappa$ such that

$$(8.3) \qquad U(\varkappa A e^{i\theta}) \leq \text{Const.}\ (B + V(2A)) , \quad |\theta| \leq \pi .$$

Proof: Following Baernstein ([2], the proof of prop. 7), we define

$$V^*(r) = \begin{cases} V(t) , & 0 \leq t \leq A , \\ AV'(A) \log (r/A) + V(A) , & t > A . \end{cases}$$

V* is a nondecreasing, convex function of log r on $(0,\infty)$. Using Poisson's formula for a semicircle (cf. Boas, p. 2), we see that if $z = re^{i\theta} = x + iy$, $x,y \in \underline{R}$,

$$G(re^{i\theta}) = (y/\pi) \int_{-A}^{0} P_1(A,z,t)\, V(|t|)\, dt +$$

$$+ (2Ry/\pi) \int_{0}^{\pi} G(Ae^{i\eta})\, P_2(A,z,\eta)\, d\eta = G_1(z) + G_2(z) ,$$

where

$$P_1(A,z,t) = (t^2 - 2tx + r^2)^{-1} - A^2(A^4 - 2tA^2 x + r^2t^2)^{-1} ,$$

$$P_2(A,z,\eta) = (A^2 - r^2) \sin \eta\, |A^2e^{2i\eta} - 2Axe^{i\eta} + r^2|^{-2} ,$$

$$G_1(z) = (y/\pi) \int_{0}^{\infty} (t^2 + 2tx + r^2)^{-1} V^*(t)\, dt ,$$

$$G_2(z) = G(z) - G_1(z) .$$

From proposition 2 in Baernstein [2] (details are given at the end of this section), we see that

$$G_{1\theta}(re^{i\theta}) = V(0)/\pi + \pi^{-1} \int_0^A \log |1 + re^{i\theta}/t| \, d\,\{t\,V'(t)\} \,.$$

Hence $G_{1\theta}$ is subharmonic in $\underline{C}$ and

$$G_{1\theta}(re^{i\theta}) \le G_{1\theta}(r) \le \text{Const.}\ (AV'(A) + V(A)) \ , \ r < A \,.$$

It is clear that G_2 can be extended to a harmonic function in $\Delta(0,A)$ and that

$$|G_2(z)| \le \text{Const.}\ \{B + V(A) + AV'(A)\} \ , \ |z| < A \,.$$

If $\varkappa$ is given, $0 < \varkappa < 1$, there exists a constant only depending on $\varkappa$ such that

$$U(\varkappa\ A\ e^{i\theta}) \le \text{Const.}\ \{B + V(A) + AV'(A)\} \,.$$

Since V is convex with respect to $\log r$ on $(0,2A)$, we obtain (8.3), and lemma 8.1 is proved.

Proof of Theorem 8.1: We use the notation preceeding formulas (2.12) and (2.12') in § 2. Let $z = re^{i\theta}$, $w = \rho e^{i\varphi}$, and assume that $z = w^{\beta/\pi}$, $|\varphi| < \pi$, where the positive real axis is mapped onto the positive real axis. We choose the functions of Lemma 8.1 in the following way. Let $A = R^{\pi/\beta}$,

$$V(\rho) = (\pi/\beta)\ v(re^{i\beta}) \ , \qquad 0 < r \ ,$$

$$G(\rho e^{i\varphi}) = (\pi/\beta)\ H(re^{i\theta}) \ , \qquad |\theta| < \beta \ , \ 0 < r < R \ ,$$

$$U(\rho e^{i\varphi}) = g(\rho e^{i\varphi}) \ , \qquad w \in \Delta(0,A) \,.$$

By Lemma 8.1, $U = g$ is subharmonic in $\Delta(0,A)$ and the Riesz mass of g is on $(-A,0)$. The convexity of V is a consequence of the convexity of $r \to v(re^{i\beta})$ which was mentioned in the summary of Baernstein's paper in § 2 (also cf. p. 76).

Furthermore, it follows from (8.3) that

$$g(R^* e^{i\varphi}) \leq \text{Const. } (1 + M(2^\gamma R)) , \quad |\varphi| \leq \pi . \tag{8.4}$$

Since (8.1) holds and $\nu(r) \leq m_\beta(r)$, we see from (2.12') and (2.13') that when $0 < \rho < A$,

$$M(\rho,g) \geq 2M(\rho^\gamma,u) , \tag{8.5}$$

$$m(\rho,g) + M(\rho,g) \leq g(-\rho) + g(\rho) \leq (1 + \cos \pi\gamma\psi(\rho^\gamma)) M(\rho,g) . \tag{8.6}$$

We can now use Theorem 5.1 with $r \to \psi(r)$ replaced by $\rho \to \gamma\psi(\rho^\gamma)$. We note in particular that the smallest value of the new step-function is $\gamma\mu$. It follows from (5.4) and (8.4) - (8.6) that

$$M(\rho^\gamma,u) \leq \text{Const.}\{|g(0)| + (1+M(2^\gamma R)) \exp \{-\gamma \int_\rho^{R^*} \Psi_\beta(t^\gamma)\, dt/t\}\} , \quad \rho < R^* .$$

Changing variables, we obtain

$$M(R) \leq \text{Const.}\{|g(0)| + (1+M(2^\gamma R))\exp \{-\int_r^{R/2} \Psi_\beta(s)\, ds/s\}\} , \quad r_0 < r < R/2 .$$

The proof of Corollary 2 is immediate from the proof of Corollary 2 of Theorem 5.1.

As a second consequence of Baernstein's theorem, we use methods similar to those of Barry [1] to prove the following new result. For convenience, we repeat certain definitions: ρ_0 and λ_0 denote the order and the lower order of a function u subharmonic in the plane, $T(r) = (2\pi)^{-1} \int_0^{2\pi} u^+ (re^{i\theta})\, d\theta$, and the upper and lower logarithmic densities $\overline{\Lambda}$ and $\underline{\Lambda}$ are defined in § 5.

Let

$$C(\lambda) = \begin{cases} \sin \pi\lambda/\pi\lambda \,, & 0 < \lambda < 1/2 \,, \\ (\pi\lambda)^{-1} \,, & \lambda \geq 1/2 \,, \end{cases}$$

$$F = \{r: T(r) > C(\lambda)\, M(r)\} \,.$$

<u>Theorem 8.2</u>: <u>Let u be subharmonic in $\underline{C}$</u> . Then

(i) $$\overline{\Lambda}\, F \geq 1 - \lambda_o/\lambda \,, \qquad \lambda > \lambda_o \,,$$

(ii) $$\underline{\Lambda}\, F \geq 1 - \rho_o/\lambda \,, \qquad \lambda > \rho_o \,.$$

<u>Remark</u>: These estimates are best possible. This can be proved using Lemma 3 in Hayman [2]. An extension of Theorem 8.2 to $\underline{R}^d$, $d > 3$, can be found in Essén and Shea [1].

<u>Proof of Theorem 8.2</u>: It is easy to see that if Theorem 8.2 is true when u^+ is continuous, it will also be true in general. We can therefore assume that u^+ is continuous. We first consider the case $\lambda \geq 1/2$.

We repeat some material from section 2. There is one change: we replace u by u^+. Consider the function $v(re^{i\theta},u^+)$ defined as in (2.11). For each fixed $re^{i\theta}$, there exists an interval $I(r,\theta)$ for which the supremum in (2.11) is attained. Let $\nu(r,\theta) = \inf u^+(re^{i\omega})$, $\omega \in I(r,\theta)$.

We introduce

$$E = \{r: (2\pi)^{-1}\, v(re^{i\pi/2\lambda}\,, u^+) \leq C(\lambda)\, M(r,u^+)\} \,,$$

We see that Compl. $E \subset F$. In the rest of the proof, the functions v and M are defined with respect to u^+ .

We start from a somewhat more precise form of Key Inequality I in Baernstein [2]. The value of the constant is due to Kjellberg ([2], p. 193, formula (23)). Then

$$(8.7)\qquad \int_r^R (\nu(t,\theta) - \cos\theta\lambda\, M(t))\, t^{-1-\lambda}\, dt >$$

$$> \lambda^{-1}(1 - \cos\theta\lambda)\, r^{-\lambda} M(r) - \text{Const.}\, R^{-\lambda} M(R)\ ,\ r < R\ ,\ \theta\lambda < \pi/2\ .$$

Since u^+ is continuous,

$$(8.8)\qquad v_\theta(re^{i\theta}) \geq 2\nu(r,\theta)\ ,\ \text{a.e.}\ ,\ 0 < \theta < \pi/2\lambda\ .$$

We use (8.8) to replace $\nu(t,\theta)$ in (8.7) by $v_\theta(te^{i\theta})/2$. Integrating with respect to θ , $0 < \theta < \pi/2\lambda$, we see that

$$(8.9)\ J(r) = r^\lambda\{\int_r^R((2\pi)^{-1}v(te^{i\pi/2\lambda}) - C(\lambda)M(t))t^{-1-\lambda}dt + \text{Const.}R^{-\lambda}M(R)\} >$$

$$> \lambda^{-1}\{(2\lambda)^{-1} - C(\lambda)\}\, M(r)\ .$$

From (8.9), we see that

$$r\,J'(r) = \lambda\,J(r) - \{(2\pi)^{-1}\,v(re^{i\pi/2\lambda}) - C(\lambda)\,M(r)\} \geq \lambda\,J(r)\ ,\ r \in E\ ,$$

$$r\,J'(r) \geq \lambda\,J(r) - \{(2\lambda)^{-1} - C(\lambda)\}\,M(r) \geq 0\ ,\ 0 < r < R\ .$$

Integrating over E , it is clear that

$$\lambda \int_{E\cap(1,R)} dt/t \leq \log\,(J(R)/\,J(1)) = \log M(R) + \text{Const.}$$

Since Compl. $E \subset F$, we see that

$$\lambda(\log R - m_\ell\, F(1,R)) \leq \log M(R) + \text{Const.}$$

Dividing by $\log R$ and letting $R \to \infty$ in the appropriate way, we obtain Theorem 8.2 in the case $\lambda \geq 1/2$.

The only difference in the case $0 < \lambda < 1/2$ is that the integration over $(0,\ \pi/2\lambda)$ is replaced by an integration over $(0,\pi)$.

I. To prove the formula at the top of p. 72, we note that

(8.10) $$\frac{\partial}{\partial\theta}\left(\frac{r\sin\theta}{r^2+t^2+2tr\cos\theta}\right) = -\frac{\partial}{\partial t}\operatorname{Re}\frac{re^{i\theta}}{t+re^{i\theta}},$$

(8.11) $$\operatorname{Re}\frac{re^{i\theta}}{t(t+re^{i\theta})} = -\frac{\partial}{\partial t}\log\left|1+\frac{re^{i\theta}}{t}\right| .$$

We know that

$$G_1(re^{i\theta}) = \pi^{-1}\int_0^\infty \frac{r\sin\theta}{r^2+t^2+2tr\cos\theta}\, v^*(t)\,dt .$$

Differentiating under the integral sign, we see from (8.10) and (8.11) that

$$G_{1\theta}(re^{i\theta}) = \pi^{-1}\int_0^\infty v^*(t)\left(-\frac{\partial}{\partial t}\operatorname{Re}\frac{re^{i\theta}}{t+re^{i\theta}}\right)dt =$$

$$= v(0)/\pi + \pi^{-1}\int_0^\infty \operatorname{Re}\frac{re^{i\theta}}{t(t+re^{i\theta})}\, tv^{*\prime}(t)\,dt =$$

$$= v(0)/\pi + \pi^{-1}\int_0^A \log\left|1+\frac{re^{i\theta}}{t}\right|\, d(tv'(t)) ,$$

and we have proved the formula.

II. At the bottom of p. 72, we use the logarithmic convexity of the function $r \to v(re^{i\beta})$. A similar result on the logarithmic convexity of a Baernstein function is stated and proved in Theorem 9.1c.

9. On two theorems of A. Baernstein

Since sections 1 - 8 were written, A. Baernstein has shown that his method can be used also to solve extremal problems for univalent functions and harmonic measures (see Baernstein [3]). Baernstein has earlier used this technique to prove that the spread conjecture of Edrei is true (see Baernstein [1]) and, as discussed in sections 2 and 8, to give an interesting generalization of the cos $\pi\lambda$-theorem of Kjellberg (see Baernstein [2]). These important applications make it worthwhile to include an account of the main results of Baernstein in these notes. The proofs given here are essentially those of Baernstein modified in a way suggested by Peter Sjögren: Lemmas 9.1 and 9.2 below are due to him. I also gratefully acknowledge interesting discussions with Christer Borell.

After having stated Lemmas 9.1 and 9.2, we give two results of Baernstein on subharmonic functions and on differences of subharmonic functions: Theorems 9.1 and 9.2. In Theorem 9.3, we extend these results to subharmonic functions in $\underline{R}^d$, $d \geq 3$. Finally, we give another result of Baernstein on an estimate of harmonic measures: Theorem 9.4. This last result is a consequence of Theorem 9.1.

We begin with the lemmas of Sjögren. Let $X = \{z: |z| = 1\}$. If $E \subset X$ and $\alpha \in \underline{R}$ is given, let $E_\alpha = e^{i\alpha}E$. The complement of E with respect to X is denoted by CE .

Lemma 9.1: Let $E \subset X$ be a measurable set such that E and CE both have positive measure. Then there exists $\delta(E) > 0$ such that

$$m(E \cap E_\eta) \leq mE - |\eta| \ , \ |\eta| < \delta(E) \ .$$

Proof: Choose $\eta > 0$. Let $a \in E$, $b \notin E$ be two Lebesgue points for the characteristic function χ_E of E . Then if $\eta < b - a$,

$$\int_a^b \chi_E(x)\ \chi_{E_\eta}(x)dx \le \int_a^{b-\eta} \chi_{E_\eta}(x)dx + \int_{b-\eta}^b \chi_E(x)dx =$$

$$= \int_a^b \chi_E(x)dx - \int_a^{a+\eta} \chi_E(x)dx + \int_{b-\eta}^b \chi_E(x)dx \ .$$

Since a and b are Lebesgue points, there exists $\delta_1 > 0$ such that if $|\eta| < \delta_1$,

$$\int_a^{a+\eta} \chi_E(x)dx \ge 9\eta/10 \ ,$$

$$\int_{b-\eta}^b \chi_E(x)dx \le \eta/10 \ .$$

If E is an interval or differs from an interval only in a null set, the lemma is trivial. Otherwise, there exist two disjoint intervals of the type described above and

$$\int_X \chi_E(x)\ \chi_{E_\eta}(x)dx \le \int_X \chi_E(x)dx - 16\eta/10 \ , \ |\eta| < \delta(E) \ .$$

The constant $\delta(E)$ can be chosen as the smallest one of the two numbers δ_1 which we obtain from the two disjoint intervals. Lemma 9.1 is proved.

Lemma 9.2: Let $E \subset X$ be as in Lemma 9.1. Let $\alpha > 0$ be given, $2\alpha < \delta(E)$, where $\delta(E)$ is defined above. Then there exist measurable sets A and B such that

(i) $A \cup B = E_{-\alpha} \cup E_\alpha$,

(ii) $A \cap B = E_{-\alpha} \cap E_\alpha$,

$mA = mE - 2\alpha$, $mB = mE + 2\alpha$.

Furthermore, for all signed Borel measures μ and ν with finite total variation, we have

(9.1) $\mu(E_{-\alpha})+\mu(E_\alpha)+\nu(CE_{-\alpha})+\nu(CE_\alpha) = \mu(A)+\mu(B)+\nu(CA)+\nu(CB)$.

Proof: Let $A_1 = E_{-\alpha} \cap E_\alpha$ and $B_1 = E_{-\alpha} \cup E_\alpha$. By Lemma 9.1, $mA_1 \leq mE - 2\alpha$. Let F be a measurable subset of $B_1 \setminus A_1$ such that $mF = mE - 2\alpha - mA_1$. Let $A = A_1 \cup F$ and $B = B_1 \setminus F$. It is easily checked that the first part of the lemma is true. To prove (9.1), let $\sigma = \mu - \nu$. Since $\nu(CE_\alpha) = \nu(X) - \nu(E_\alpha)$, (9.1) is equivalent to $\sigma(E_{-\alpha}) + \sigma(E_\alpha) = \sigma(A) + \sigma(B)$, which is obvious. Lemma 9.2 is proved.

We can now discuss the theory of A. Baernstein. Let u be subharmonic in $\Delta(0,R)$, where $R > 0$ is given. Let, if $0 < r < R$ and $0 \leq \theta \leq \pi$,

$$u^*(re^{i\theta}) = \sup_E \int_E u(re^{i\omega})d\omega ,$$

where the supremum is taken over all measurable sets $E \subset (-\pi,\pi]$ such that $mE = 2\theta$.

Example: Let $f(z) = \prod_1^\infty (1 + z/a_\nu)$, where $a_\nu > 0$, $\nu = 1,2,\ldots$. Prove that if $u(z) = \log|f(z)|$, then $u^*(re^{i\theta}) = \int_{-\theta}^{\theta} u(re^{i\omega})d\omega$ is harmonic in the upper halfplane.

Remark: In section 1, we used the notation u* for "an associated subharmonic function" which is different from the Baernstein function u*. In this section, u* is always used in the sense of Baernstein.

The main result on subharmonic functions given by Baernstein is (cf. Baernstein [2], § 2; [3]):

Theorem 9.1: Let u be subharmonic in $\Delta(0,R)$. Then

a) $u^*(re^{i\theta})$ is continuous in $\{z: \operatorname{Im} z \geq 0 , 0 < |z| < R\}$.

b) $u^*(re^{i\theta})$ is subharmonic in $\{z: \operatorname{Im} z > 0\} \cap \Delta(0,R)$.

c) If β is given, $0 \leq \beta \leq \pi$, $u^*(re^{i\beta})$ is a convex function of $\log r$, $0 < r < R$.

d) If $re^{i\theta}$ is given, $0 \leq \theta \leq \pi$, there exists a measurable set $E = E(r,\theta)$ such that

$$u^*(re^{i\theta}) = \int_E u(re^{i\omega})d\omega \ , \ 0 < r < R \ .$$

Proof: We start with d). Since $\theta \to u(re^{i\theta})$ is a measurable function, we can consider $A(t) = \text{meas.}\ \{\theta: u(re^{i\theta}) \geq t\}$ which is a decreasing function of t on $\underline{R}$. It is now easy to find a set $E(r,\theta)$, as stated in d).

We now turn to a). Let R_1 be given, $0 < R_1 < R$. In $\overline{\Delta(0,R_1)}$, we construct a decreasing sequence of continuous subharmonic functions $\{u_n\}_1^\infty$ such that $u_n \to u$, $n \to \infty$. Consider

$$I_n(r) = \int_{-\pi}^{\pi} u_n(re^{i\theta})d\theta \to \int_{-\pi}^{\pi} u(re^{i\theta})d\theta = I(r) \ , \ n \to \infty \ .$$

Since $I_n(r)$ and $I(r)$ are convex with respect to $\log r$, they are certainly continuous on $(0,R)$. Using Dini's theorem, we see that $I_n(r) \to I(r)$ uniformly on each interval $[\delta,R_1]$, where $\delta > 0$ is arbitrary. Using d), we find a measurable set $E_n = E_n(r,\theta)$ such that $mE_n = 2\theta$ and such that $u_n^*(re^{i\theta}) = \int_{E_n} u_n(re^{i\omega})d\omega$. Then

$$0 \leq u_n^*(re^{i\theta}) - u^*(re^{i\theta}) \leq \int_{E_n} (u_n-u)(re^{i\omega})d\omega \leq I_n(r) - I(r) \ .$$

Hence $u_n^* \to u^*$ uniformly in $\{z: \text{Im} z \geq 0 \ , \ 0 < \delta \leq |z| \leq R_1\}$. Since $\{u_n^*\}$ is a sequence of continuous functions, we have proved a).

To prove b), let $re^{i\theta}$ be given, $0 < \theta < \pi$, $0 < r < R$ and let $E = E(r,\theta)$ be the measurable set as in d). From the subharmonicity of u, we see that

$$u^*(re^{i\theta}) = \int_E u(re^{i\omega})d\omega \leq \frac{1}{2\pi} \int_E d\omega \int_{-\pi}^{\pi} u(e^{i\omega}(r + \rho e^{i\varphi}))d\varphi \ .$$

Let us put $r + \rho e^{i\varphi} = e^{i\alpha(\varphi)}|r + \rho e^{i\varphi}|$, where $\alpha(\varphi) \geq 0$ when $0 \leq \varphi \leq \pi$ We note that $\alpha(-\varphi) = -\alpha(\varphi)$. It is clear that when ρ is small compared to r , $\alpha(\varphi)$ can be chosen close to 0. Hence the last integral can be written as

$$\frac{1}{2\pi}\int_E d\omega \int_0^{\pi} \{u(e^{i\omega+i\alpha(\varphi)}|r+\rho e^{i\varphi}|) + u(e^{i\omega-i\alpha(\varphi)}|r+\rho e^{i\varphi}|)\}\, d\varphi =$$

$$= \frac{1}{2\pi}\int_0^{\pi} d\varphi\, \{\int_{E_\alpha} + \int_{E_{-\alpha}}\}\, u(e^{i\omega}|r+\rho e^{i\varphi}|)\, d\omega\ .$$

Applying Lemma 9.2, this integral is equal to

$$\frac{1}{2\pi}\int_0^{\pi} d\varphi\, \{\int_A + \int_B\}\, u(e^{i\omega}|r+\rho e^{i\varphi}|)\, d\omega \le$$

$$\le \frac{1}{2\pi}\int_0^{\pi} \{u^*(|r+\rho e^{i\varphi}|e^{i(\theta-\alpha(\varphi))}) + u^*(|r+\rho e^{i\varphi}|e^{i(\theta+\alpha(\varphi))})\}\, d\varphi =$$

$$= \frac{1}{2\pi}\int_{-\pi}^{\pi} u^*(e^{i\theta}(r+\rho e^{i\varphi}))\, d\varphi\ .$$

Here $A = A(\varphi)$ and $B = B(\varphi)$ are the two sets defined in Lemma 9.2 starting from E and $\alpha = \alpha(\varphi)$. We remind the reader that $mA = 2(\theta-\alpha)$ and $mB = 2(\theta+\alpha)$.

We have proved that u^* has the local subharmonic mean value property. Since u^* is continuous, u^* is subharmonic.

It remains to prove c). Let $r_0 e^{i\beta}$ be given, let $E = E(r_0,\beta)$ and consider $Q(re^{i\varphi}) = \int_E u(re^{i(\omega+\varphi)})\, d\omega$. We note that Q is subharmonic in the plane, that $M(r,Q) = \sup_\varphi Q(re^{i\varphi})$ is a convex function of log r and that $u^*(r_0 e^{i\beta}) = M(r_0,Q)$. Since $M(r,Q) \le u^*(re^{i\beta})$, we see that for appropriate real values of s ,

$$u^*(r_0 e^{i\beta}) \le \frac{1}{2}\, \{M(sr_0,Q) + M(r_0/s,Q\} \le \frac{1}{2}\, \{u^*(sr_0 e^{i\beta}) + u^*((r_0/s)e^{i\beta})\}\ .$$

Hence $r \to u^*(re^{i\beta})$ is a convex function of log r and c) is proved. The proof of Theorem 9.1 is complete.

As mentioned in § 3 in Baernstein [3], results of this type are true also for differences of subharmonic functions. Let v and w be two functions which are subharmonic in a disk $\Delta(0,R)$ and let $u = v - w$. Let μ be the Riesz mass of w. For simplicity, we assume that v and w are harmonic near the origin. We introduce $N(r,u) = \int_0^r \mu\{|z| < t\}dt/t$. (If u is the logarithm of a meromorphic function, $N(r,u) = \int_0^r n(t)dt/t$ where n(r) is the number of poles in $\Delta(0,r)$.) We also introduce

$$m^*(re^{i\theta}) = \sup_E \int_E u(re^{i\omega})d\omega ,$$

where the supremum is taken over all measurable sets $E \subset (-\pi,\pi]$ such that $mE = 2\theta$. Let

$$T^*(re^{i\theta}) = m^*(re^{i\theta}) + 2\pi N(r,u) .$$

Example: Let $f(z) = \Pi(1+z/a_\nu)$ and $g(z) = \Pi(1-z/b_\nu)$, where a_ν and b_ν are positive, $\nu = 1,2...$. Let $v(z) = \log|f(z)|$, $w(z) = \log|g(z)|$ and put $u = v - w$. Prove that $m^*(re^{i\theta}) = \int_{-\theta}^{\theta} u(re^{i\omega})d\omega$ and that $T^*(re^{i\theta})$ is harmonic in the upper halfplane.

The following result can be found in Baernstein ([1], [3]).

Theorem 9.2: Let u be the difference of two subharmonic functions in $\Delta(0,R)$. Then the conclusions of Theorem 9.1 a) - c) are true with u^* replaced by T^*.

Proof: As in the previous proof, it is easy to see that if $re^{i\theta}$ is given, $0 \leq \theta \leq \pi$, there exists a measurable set $E = E(r,\theta)$ such that

$$m^*(re^{i\theta}) = \int_E u(re^{i\omega})\, d\omega . \tag{9.2}$$

Let $F = CE$. Then

$$m^*(re^{i\theta}) = \int_E v(re^{i\omega})d\omega + \int_F w(re^{i\omega})d\omega - \int_0^{2\pi} w(re^{i\omega})d\omega .$$

Using the well-known formula

$$\int_0^{2\pi} \log\left|1 - \frac{re^{i\omega}}{\rho}\right| d\omega = 2\pi \log^+(r/\rho) ,$$

we see that

$$\int_0^{2\pi} w(re^{i\omega})d\omega = \int_{\Delta(0,R)} d\mu(\rho)\log^+(r/\rho) = 2\pi\, N(r,u) .$$

Hence

$$T^*(re^{i\theta}) = \int_{E(r,\theta)} v(re^{i\omega})d\omega + \int_{F(r,\theta)} w(re^{i\omega})d\omega .$$

In particular we see that as an alternative definition of T^*, we can take

$$(9.3) \qquad T^*(re^{i\theta}) = \sup_E \{\int_E v(re^{i\omega})d\omega + \int_F w(re^{i\omega})d\omega\} ,$$

where $F = CE$ and the supremum is taken over all measurable sets $E \subset (-\pi,\pi]$ such that $mE = 2\theta$.

To prove a), we use (9.3) and approximate v and w by decreasing sequences of continuous subharmonic functions just in the proof of Theorem 9.1.

To prove b), let $re^{i\theta}$ be given and let E be an associated measurable set of measure 2θ according to (9.2). Let us also this time put $r + \rho e^{i\varphi} = |r + \rho e^{i\varphi}|e^{i\alpha(\varphi)}$. The argument is very similar to the one used in the previous proof: the only difference is that this time, we shall also need (9.1) from Lemma 9.2. The proof goes as follows:

$$T^*(re^{i\theta}) = \int_{E(r,\theta)} v(re^{i\omega})d\omega + \int_{F(r,\theta)} w(re^{i\omega})d\omega \leq$$

$$\leq \int_E (2\pi)^{-1} \int_{-\pi}^{\pi} v(e^{i\omega}(r+\rho e^{i\varphi}))d\varphi\, d\omega + \int_F (2\pi)^{-1} \int_{-\pi}^{\pi} w(e^{i\omega}(r+\rho e^{i\varphi}))d\varphi\, d\omega =$$

$$= (2\pi)^{-1}\{\int_0^{\pi} d\varphi \{(\int_{E_\alpha} + \int_{E_{-\alpha}})v(e^{i\omega}|r+\rho e^{i\varphi}|)d\omega + (\int_{CE_\alpha} + \int_{CE_{-\alpha}})w(e^{i\omega}|r+\rho e^{i\varphi}|)d\omega\}\} =$$

$$= (2\pi)^{-1}\{\int_0^{\pi} d\varphi \{(\int_A + \int_B)v(e^{i\omega}|r+\rho e^{i\varphi}|)d\omega + (\int_{CA} + \int_{CB})w(e^{i\omega}|r+\rho e^{i\varphi}|)d\omega\}\} \leq$$

$$\leq (2\pi)^{-1} \{\int_0^{\pi} d\varphi \{T^*(|r+\rho e^{i\varphi}|e^{i(\theta-\alpha)}) + T^*(|r+\rho e^{i\varphi}|e^{i(\theta+\alpha)})\}\} =$$

$$= (2\pi)^{-1} \int_{-\pi}^{\pi} T^*(e^{i\theta}(r+\rho e^{i\varphi}))d\varphi \,.$$

We remind the reader that $F = CE$, that $mE = 2\theta$, that A and B are defined as in Lemma 9.2 and thus that $mA = 2(\theta-\alpha)$ and that $mE = 2(\theta+\alpha)$. We have proved b).

To prove c), let $r_o e^{i\beta}$ be given and let $E = =(r_o,\beta)$ be a maximizing set in (9.3). The convexity of $T^*(re^{i\beta})$ with respect to $\log r$ follows by considering the subharmonic function

$$Q(re^{i\varphi}) = \int_E v(re^{i(\omega+\varphi)})d\omega + \int_{CE} w(re^{i(\omega+\varphi)})d\omega$$

in the same way as in the previous proof. We have proved Theorem 9.2.

We would also like to discuss how the results of Baernstein can be extended to higher dimensions. In $\underline{R}^d$, $d \geq 3$, let

$x = (x_1,\dots x_{d-1},t) = (x',t)$. Let Ω be an open connected set in $\underline{R}^{d-1}$.
We shall discuss two cases:

a) u is a nonnegative subharmonic function in $\Omega \times \underline{R}$ which is such that if $x' \in \Omega$ is given, $u(x',t)$ is 0 outside a compact set.

b) u is a subharmonic function in $\Omega \times \underline{R}$ such that $u(x',t+2T) = u(x',t)$ for all $(x',t) \in \Omega \times \underline{R}$.

We define

$$u^*(x',t) = \sup_E \int_E u(x',s)ds \, , \tag{9.4}$$

where in case a) the supremum is taken over all measurable sets E of measure 2t ; u^* is defined on $\Omega \times [0,\infty)$. In case b) the supremum is taken over all measurable sets $E \subset [0,2T]$ of measure 2t ; u^* is defined on $\Omega \times [0,T]$.

Theorem 9.3: Let u be as in a) or b). Then

i) $u^*(x',t)$ is subharmonic in the interior of its domain of definition.

ii) If t is fixed, $x' \to u^*(x',t)$ is subharmonic in Ω .

We shall prove that $\Delta u^* \geq 0$, where the derivatives are taken in the distributional sense. It follows that u^* is subharmonic. The same method will work if we want to consider differences of subharmonic functions and thus generalize Theorem 9.2 to the situation considered here.

The first part of Theorem 9.3 could also have been obtained by applying Lemmas 9.1 and 9.2 once more (see the following paper of Christer Borell!). On the other hand, we could also have proved Theorems 9.1 and 9.2 by showing that $\Delta u^* \geq 0$.

In the proof, we want to work with functions which are subharmonic in all of $\underline{R}^d$. For this purpose, we need a lemma.

Lemma 9.3: Let u be as above, let $x'_0 \in \Omega$ and let S_1, S_2 and S_3 be (d-1)-dimensional balls centered at x_0, contained in Ω and with radii r_1, r_2 and r_3 where $0 < r_1 < r_2 < r_3$. Then there exists a function $\tilde{u}$ which is subharmonic in $\underline{R}^d$ and is such that,

$$\tilde{u}(x) = u(x) \quad , \quad x \in \bar{S}_1 \times \underline{R} . \tag{9.5}$$

There exist constants A and B such that

$$\tilde{u}(x) = \begin{cases} A|x' - x'_0|^{3-d} + B \,, & |x' - x'_0| \geq r_3 \,, \ d > 3 \\ \\ A \log|x' - x'_0| + B \,, & |x' - x'_0| \geq r_3 \,, \ d = 3 \,. \end{cases} \tag{9.6}$$

Proof: In $(S_3 \setminus S_1) \times \underline{R}$, we replace u by its least harmonic majorant and obtain a subharmonic function u_1 which is u in $S_1 \times \underline{R}$ and harmonic in $(S_3 \setminus S_1) \times \underline{R}$. In particular, u_1 is continuous and bounded on $\partial S_2 \times \underline{R}$. Now choose two numbers A_1 and B_1 such that

$$A_1 > \sup u_1(x) \quad , \quad x \in \partial S_3 \times \underline{R} \,,$$

$$A_2 < \inf u_1(x) \quad , \quad x \in \partial S_2 \times \underline{R} \,.$$

If $d \geq 4$, let $\beta(x) = A|x' - x'_0|^{3-d} + B$, where A and B are chosen so that

$$\beta(x) = A_1 \quad , \quad x \in \partial S_3 \times \underline{R} \,,$$

$$\beta(x) = A_2 \quad , \quad x \in \partial S_2 \times \underline{R} \,.$$

It is clear that β is a harmonic function. We define $\tilde{u}$ in the following way:

$$\tilde{u}(x) = \begin{cases} \beta(x) \,, & x \notin S_3 \times \underline{R} \,, \\ \operatorname{Max}(\beta(x) \,, u_1(x)) \,, & x \in (S_3 \setminus S_2) \times \underline{R} \,, \\ u_1(x) \,, & x \in S_2 \times \underline{R} \,. \end{cases}$$

It is easily checked that $\tilde{u}$ has the properties stated in the lemma. If $d = 3$, a similar argument is used. Lemma 9.3 is proved.

Proof of Theorem 9.3: Let $S \subset \Omega$ be a (d-1)-dimensional ball. It is sufficient to prove that the theorem is true if u is subharmonic on any set of the type $\bar{S} \times \underline{R}$. Due to Lemma 9.3, we can assume that u is subharmonic in all of $\underline{R}^d$ and small at infinity as stated in (9.6).

We shall use real-analytic functions (cf. Rudin 6.2.1). Let $k(x) = C(d) \exp\{-|x|^2\}$, where C_d is chosen so that $\int_{\underline{R}^d} k(x)dx = 1$. Let, if $\varepsilon > 0$ is given, $k_\varepsilon(x) = \varepsilon^{-d} k(x/\varepsilon)$, and let

$$(9.7) \quad u_\varepsilon(x) = \int_{\underline{R}^d} k_\varepsilon(x-y)\, u(y)dy = C(d) \int_0^\infty \rho^{d-1} \exp(-\rho^2) \{\int_{|y|=1} u(x+\varepsilon\rho y) d\sigma(y)\} d\rho ,$$

where σ is the uniform Lebesgue measure on the unit sphere in $\underline{R}^d$ with total mass 1. It is easy to see that $\{u_{1/n}\}_{n=1}^\infty$ is a decreasing sequence of real-analytic functions which tends to u.

If Theorem 9.3 is true for real-analytic functions, it will also be true for decreasing sequences of real-analytic functions, i.e., it will be true in general.

It remains to prove Theorem 9.3 assuming that u is real-analytic. We note that in case a), if $x' \in S$ is given, the support of $t \to u(x',t)$ need no longer be a compact set. On the other hand, it is clear that $u(x',t) \to 0$, $|t| \to \infty$, uniformly in $\bar{S}$. (We assume that u is a convolution of a kernel and a nonnegative function as in (9.7)). Hence in case a, definition (9.4) is meaningful also for our approximating real-analytic functions. In case b, no discussion of this type is necessary.

Let $x' \in S$ be given. Then $t \to u(x',t)$ is a real-analytic function in t. (In the sequel, we assume that this function is non-constant. We leave the modifications when $t \to u(x',t)$ is constant to the reader.) Let $E = E(x',t)$ be a maximizing set in (9.4) of measure 2t. Then there exists $c = c(x',t) \in \underline{R}$ such that in case a and b, respectively,

(9.8a) $$E(x',t) = \{s\colon u(x',s) > c\} ,$$

(9.8b) $$E(x',t) = \{s\colon u(x',s) > c , s_0 \leq s \leq s_0 + 2T\} ,$$

where s_0 is such that $u(x',s_0) \leq c$.

Let $E(x',t) = \cup(a_\nu,b_\nu)$. Since u is real-analytic, this union contains only a finite number of intervals. If (a,b) is one of these intervals, we see from (9.8) that

(9.9) $$u_t'(x',b) - u_t'(x',a) \leq 0 .$$

Since u is real-analytic, u^* is a continuous function. To prove that $\Delta u^* \geq 0$, we shall consider symmetric differences. Let, if $h > 0$ is given,

$$\Delta_t^2(u,h)(x',t) = h^{-2} \{u(x',t+h) + u(x',t-h) - 2u(x',t)\} ,$$

$$\Delta_{x'}^2(u,h)(x',t) = \sum_1^{d-1} h^{-2} \{u(x'+he_k,t) + u(x'-he_k,t) - u(x',t)\} .$$

Here e_k is the unit vector along the x_k-axis. Let (a,b) be one of the intervals in $E(x',t)$. Let

$$E_1(x',t+h) = (a-h,b+h) \cup \{E\setminus(a,b)\} ,$$

$$E_1(x',t-h) = (a+h,b-h) \cup \{E\setminus(a,b)\} .$$

If h is sufficiently small, the extended interval (a-h,b+h) is disjoint from all intervals in $E(x',t)$ except (a,b). To compute Δu^* , we first note that

$$\Delta_t^2(u^*,h)(x',t) = h^{-2}\{ \int_{E(x',t+h)} + \int_{E(x',t-h)} - 2\int_{E(x',t)} \} u(x',s)ds \geq$$

$$\geq h^{-2}\{ \int_{E_1(x',t+h)} + \int_{E_1(x',t-h)} - 2\int_{E(x',t)} \} u(x',s)ds =$$

$$= h^{-2}\{ \int_{a-h}^{b+h} + \int_{a+h}^{b-h} - 2\int_{a}^{b} \} u(x',s)ds \to u_t'(x',b) - u_t'(x',a) .$$

To compute $\Delta_{x'}^2(u^*,h)(x',t)$, we note that

$$u^*(x'+he_k,t) = \int_{E(x'+he_k,t)} u(x'+he_k,s)ds \geq \int_{E(x',t)} u(x'+he_k,s)ds .$$

There is a similar formula for $E(x'-he_k,t)$. Hence

(9.10) $$\Delta_{x'}^2(u^*,h)(x',t) \geq \int_{E(x',t)} \Delta_{x'}^2(u,h)(x',s)ds \to$$

$$\to \int_{E(x',t)} \sum_{k=1}^{d-1} \frac{\partial^2 u}{\partial x_k^2}(x',s)ds \geq \int_{E(x',t)} \left(-\frac{\partial^2 u}{\partial s^2}(x',s)\right)ds =$$

$$= -\sum_{\nu} (u_t'(x',b_\nu) - u_t'(x',a_\nu)) .$$

Here we have used the fact that $\Delta u \geq 0$. Hence $\Delta(u^*,h)(x't) = (\Delta_t^2(u^*,h) + \Delta_x^2(u^*,h))(x',t)$ is bounded from below by a quantity which tends boundedly to $-\Sigma'(u_t'(x',b_\nu) - u_t'(x',a_\nu))$ as $h \to 0+$. Here Σ' denotes that we sum over all intervals in $E(x',t)$ except the one used in the estimate of $\Delta_t^2(u^*,h)$. This lower bound $-\Sigma'$ is nonnegative according to (9.9). Let $\varphi \in C^\infty$ be nonnegative and have compact support. Then

$$\int_{\underline{R}^d} u^*(x)\Delta\varphi(x)dx = \lim_{h\to 0}\int_{\underline{R}^d} u^*(x)\Delta(\varphi,h)(x)dx = \lim_{h\to 0}\int_{\underline{R}^d} \Delta(u^*,h)(x)\varphi(x)dx \geq 0 .$$

Hence $\Delta u^* \geq 0$ and we have proved Theorem 9.3 i). To prove Theorem 9.3 ii), we note that according to (9.9) and (9.10), $\Delta^2_{x'}(u^*,h)(x',t)$ is bounded from below in a similar way and thus that $\sum_{k=1}^{d-1} \frac{\partial^2 u^*}{\partial x_k^2} \geq 0$. Since $x' \to u^*(x',t)$ is continuous if t is fixed, the proof is complete.

As an application of Theorem 9.1, we shall give the proof of the following result of Baernstein (cf. [3], Theorem 3). Let D be a region in the unit disk Δ such that $\gamma = \partial D \cap \{|z| = 1\}$ is such that $\int_\gamma d\omega > 0$. Furthermore, we assume that the harmonic measure $\omega(z,\gamma,D)$ of γ with respect to D exists. Let

$$u(z) = \begin{cases} \omega(z,\gamma,D) , & z \in D , \\ 0 , & z \in \Delta\setminus D . \end{cases}$$

We note that

$$u(re^{i\theta}) \to \chi_\gamma(\theta) \text{ weakly } , r \to 1 , \tag{9.11}$$

where χ_γ is the characteristic function of γ . Let D^* be the circular symmetrization of D, let $\gamma^* = \partial D^* \cap \{|z| = 1\}$ and let $\omega(z,\gamma^*,D^*)$ be the corresponding harmonic measure. We also introduce

$$v(z) = \begin{cases} \omega(z,\gamma^*,D^*) , & z \in D^* , \\ 0 , & z \in \Delta\setminus D^* . \end{cases}$$

It is obvious that u and v are subharmonic in Δ .

Theorem 9.4: Let B be a convex, non-decreasing function on $[0,\infty)$. Then for $0 < r < 1$,

$$(9.12)\qquad \int_0^{2\pi} B(u(re^{i\theta}))d\theta \le \int_0^{2\pi} B(v(re^{i\theta}))d\theta \ .$$

Corollary: $\sup_\theta \omega(re^{i\theta},\gamma,D) \le \omega(r,\gamma^*,D^*) \quad , \quad 0 < r < 1$.

To prove the corollary, we choose $B(x) = x^p$ in (9.12), take p^{th} roots and let $p \to \infty$. To prove the theorem, we first note that it is sufficient to prove that for each real number α ,

$$(9.13)\qquad \int_0^{2\pi} (u(re^{i\theta}) + \alpha)^+ \, d\theta \le \int_0^{2\pi} (v(re^{i\theta}) + \alpha)^+ \, d\theta \quad , \quad 0 < r < 1 \ .$$

This is clear, since the convex increasing function B can be approximated by sums of the type $c + \Sigma\, b_\nu(x + \alpha_\nu)^+$, where $\{b_\nu\}$ is a sequence of positive numbers and c is a constant.

We claim that if

$$(9.14)\qquad u^*(re^{i\theta}) \le v^*(re^{i\theta}) \quad , \quad 0 < r < 1 \quad , \quad 0 < \theta < \pi \ ,$$

(9.13) will be true. This is proved in the following way. Let $E = E(r)$ be a set, $mE = 2\theta_0$, such that

$$\int_0^{2\pi} (u(re^{i\theta}) + \alpha)^+ d\theta = \int_E (u(re^{i\theta}) + \alpha)d\theta \le v^*(re^{i\theta_0}) + 2\alpha\theta_0 =$$

$$= \int_{E_1} (v(re^{i\theta}) + \alpha)d\theta \le \int_0^{2\pi} (v(re^{i\theta}) + \alpha)^+ d\theta \ .$$

Here $mE_1 = 2\theta_0$.

Hence it suffices to prove (9.14). We start with a lemma.

Lemma 9.2: $v^*(re^{i\theta}) = 2\int_0^\theta v(re^{i\omega})d\omega$, $0 \leq \theta \leq \pi$, $0 < r < 1$.

In particular, v^* is harmonic in $D_1 = D^* \cap \{\text{Im}z > 0\}$.

Proof: It is clear that $v(re^{i\theta}) = v(re^{-i\theta})$. Let

$V(re^{i\theta}) = 2\int_0^\theta v(re^{i\varphi})d\varphi$, $z \in D_1$. Since V is harmonic in D_1, $v^* - V$ is subharmonic in D_1 . It is easy to check that $v^* - V \leq 0$ on ∂D_1 . Hence $v^* \leq V$ in D_1 and we must have $v^* = V$.

Remark: We have in fact proved that $\theta \to \omega(re^{i\theta},\gamma^*,D^*)$ is decreasing when $\theta > 0$.

We can now prove (9.14). Let $\varepsilon > 0$ be given. We claim that

$$(9.15) \qquad ((u-\varepsilon)^* - v^*)(z) \leq 0 \quad , \quad z \in D^* .$$

To prove (9.15), we first use (9.11) to obtain

$((u-\varepsilon)^* - v^*)(e^{i\theta}) \leq 0$, $e^{i\theta} \in \gamma^* \cap \{\text{Im}z > 0\}$.

Let $2\theta(r)$ be the angular measure of $D^* \cap \{|z| = r\}$. Since $\int_\gamma d\varphi > 0$, we know that $\delta = \liminf_{r \to 1} \theta(r) > 0$. Hence

$\limsup_{r \to 1} \{(u-\varepsilon)^* - v^*\}(re^{i\theta(r)}) =$

$= \limsup_{r \to 1} \{ \int_{-\pi}^{\pi} (u(e^{i\varphi}) - v(e^{i\varphi}))d\varphi - 2\varepsilon\theta(r)\} \leq -2\varepsilon\delta$, and $(u-\varepsilon)^* - v^* \leq 0$ on γ^* and on that part of $\partial D_1 \cap \{\text{Im}z \geq 0\}$ which is close to the unit circle. From the definition of the $*$-functions, we see that this inequality is also true on the positive real axis. If (9.15) is false, it follows from the maximum principle that the subharmonic function $(u-\varepsilon)^* - v^*$ assumes its maximum M over $\overline{D}_1$ on ∂D_1 at $P_o = r_o e^{i\theta_o}$, where $0 < \theta_o \leq \pi$ and $0 \leq r_o < 1$. If $\inf_\theta u(r_o e^{i\theta}) = 0$, we have

$$\lim_{\theta\to\theta_0^-} \frac{\partial}{\partial\theta} ((u-\varepsilon)^* - v^*)(r_0 e^{i\theta}) = -2\varepsilon < 0 ,$$

which shows that there exist points in D_1 close to P_0 where $(u-\varepsilon)^* - v^*$ is even larger than at P_0 . If $\inf_\theta u(r_0 e^{i\theta}) > 0$, we know that u and v are both harmonic in an open annulus or in an open disk containing $C(0,r_0)$ in its interior and that $\theta_0 = \pi$. In the first case, let $\{z = 0 \leq a < |z| < b\}$ be the largest annulus of this type. Thus

$$(9.16) \qquad (u-\varepsilon)^* - v^*)(-r) = \int_{-\pi}^{\pi} u(re^{i\theta})d\theta - \int_{-\pi}^{\pi} v(re^{i\theta})d\theta - 2\varepsilon\pi$$

is a linear function of log r in $a < r < b$ (cf. Heins p. 57, formula (2.7)) and $\sup_{a\leq r\leq b} ((u-\varepsilon)^* - v^*)(-r) = M$ is assumed at a or b. But $\inf_\theta u(ae^{i\theta}) = \inf_\theta u(be^{i\theta}) = 0$. From the previous discussion, we know that the maximum can not be assumed at -a or -b . In the second case, u and v are harmonic in a disk $\Delta(0,b)$ containing $C(0,r_0)$ in its interior: we chose b so that $\Delta(0,b)$ is the largest disk of this type. In this case, the circular mean in (9.16) is constant on $0 \leq r \leq b$. Thus the maximum is assumed at -b which is impossible. The contradiction shows that (9.15) holds. Since $\varepsilon > 0$ is arbitrary, we obtain (9.14). We have proved Theorem 9.4.

Remark: When the region D is simply connected, the corollary of Theorem 9.4 is a known result: see Haliste ([1], Theorem 4.1) (in higher dimensions, we discussed related results of Haliste at the end of section 5). It has been pointed out to me by J. Krzyz that Haliste's result also follows from estimates of Green's function due to Krzyz [1]. This observation will appear in Krzyz [2].

REFERENCES

Ahlfors, L. and Heins, M.

1. Questions of regularity connected with the Phragmén-Lindelöf principle. Ann. of Math. 50 (1949), 341-346.

Azarin, V.

1. Generalization of a theorem of Hayman on subharmonic functions in an n-dimensional cone, AMS Transl. (2) 80 (1969), 119-138. Mat. Sb. 66 (108), (1965), 248-264.

Baernstein, A.

1. Proof of Edrei's spread conjecture. Proc. London Math. Soc. (3) 26, 1973, 418-434.
2. A generalization of the cos $\pi\rho$-theorem. Trans. Amer. Math. Soc. 193 (1974), 181-197.
3. Some extremal problems for univalent functions, harmonic measures and subharmonic functions. Proceedings of the conference on Classical Function Theory, Canterbury July 1973. IMS Lecture Note Series 12, 11-15.

Barry, P.D.

1. Some theorems related to the cos $\pi\rho$-theorem, Proc. London Math. Soc. XXI (1970), 334-360.

Bebernes, J.W. and Jackson, L.K.

1. Infinite interval boundary value problems for $y'' = f(x,y)$, Duke Math. Journ. 34 (1967), 39-47.

Carleman, T.

1. Sur une inégalité différentielle dans la théorie des fonctions analytiques. C.R. Acad. Sci. Paris, 196 (1933), 995-997.

Dahlberg, B.

1. Mean values of subharmonic functions, Ark. Mat. 10 (1972), 293-309.

Drasin, D.

1. Tauberian theorems and slowly varying functions, Trans. Amer. Math. Soc. 133 (1968), 333-356.

Drasin, D. and Shea, D.

1. Asymptotic properties of entire functions extremal for the cos $\pi\rho$-theorem, Bull. Amer. Math. Soc. 75 (1969), 119-122.

Edrei, A.

1. A local form of the Phragmén-Lindelöf indicator, Mathematika 17 (1970) 149-172.

Essén, M.

1. Note on "A theorem on the minimum modulus of entire functions" by Kjellberg. Math. Scand. 12 (1963), 12-14.
2. A generalization of the Ahlfors-Heins theorem.
   a) Bull. Amer. Math. Soc. 75 (1969), 127-131.
   b) Trans. Amer. Math. Soc. 142 (1969), 331-344.
3. Theorems of (λ,ρ)-type on the growth of subharmonic functions Report 1971-1, Department of Mathematics, Royal Institute of Technology, S-10044 Stockholm, Sweden.

Essén, M. and Jackson, H.

1. A comparison between minimally thin sets and generalized Azarin sets. To appear, Canad. Math. Bull. (also cf. C.R. Acad. Sci. Paris, t. 277, Série A (1973), 241-242).

Essén, M. and Lewis, J.

1. The generalized Ahlfors-Heins Theorem in certain d-dimensional cones.
   a) Proc. of the conf. on Diff. Equations in Dundee, March 1972 (Springer).
   b) Math. Scand. 33 (1973), 113-129.

Essén, M. and Shea, D.F.

1. Applications of Denjoy integral inequalities to growth problems for subharmonic and meromorphic functions. Proc. of the conference on Classical Function Theory, Canterbury, July 1973, LMS Lecture Note Series 12, 59-68.

Fuchs, W.H.J.

1. Topics in Nevanlinna theory, Proc. of the NRL Conference on Classical Function Theory, edited by F.Gross, Washington 1970.

Haliste, K.

1. Estimates of harmonic measures. Ark. Mat. 6 (1965), 1-31.

Hayman, W.

1. Questions of regularity connected with the Phragmén-Lindelöf principle. J. Math. Pures Appl. (9) 35 (1956), 115-126.
2. Some examples related to the cos $\pi\rho$ theorem. MacIntyre Memorial Volume, Ohio University Press, 1970.

Heins, M.

1. Entire functions with bounded minimum modulus; subharmonic analogues. Ann. of Math. (2) 49 (1948), 200-213.
2. On a notion of convexity connected with a method of Carleman. J. Analyse Math. 7 (1960), 53-77.

Hellsten, U., Kjellberg, B. and Norstad, F.

1. Subharmonic functions in a circle. Ark. Mat. 8 (1970), 185-193.

Kjellberg, B.

1. On certain integral and harmonic functions. Thesis, Uppsala 1948.
2. On the minimum modulus of entire functions of order less than one, Math. Scand. 8 (1960), 189-197.
3. A theorem on the minimum modulus of entire functions, Math. Scand. 12 (1963), 5-11.

Kneser, A.

1. Untersuchung und asymptotische Darstellung der Integrale gewisser Differentialgleichungen bei grossen Werthen des Arguments, Journal für die reine und angewandte Mathematik 116 (1896), 178-212.

Krzyz, J.

1. Circular symmetrization and Green's function. Bull. Acad. Polon. Sci. Sér. Math. Astronom. Phys. VII, no. 6, 1959, 327-330.

2. Lectures at the Univ. of Maryland. To appear, Springer Lecture Notes.

Lelong-Ferrand, J.

1. Etude au voisinage de la frontière des fonctions surharmoniques positives dans un demi-espace. Ann. Sci. École Norm. Sup. 66 (1949), 125-159.

Lewis, J.

1. Some theorems on the cos $\pi\lambda$-inequality. Trans. Amer. Math. Soc. 167 (1972), 171-189.

2. Subharmonic functions in certain regions. Trans. Amer. Math. Soc. 167 (1972), 191-201.

3. A note on Essén's generalization of the Ahlfors-Heins theorem, Trans. Amer. Math. Soc. 172 (1972), 339-345.

4. Extremal analytic functions in the unit circle, Ark. Mat. 10 (1972), 173-194.

5. A potential theory problem in three space. Univ. of Kentucky 1972.

Norstad, F.

1. Konvexitet hos medelvärdet av vissa subharmoniska funktioner (Convexity of the mean-value of certain subharmonic functions) Manuscript, 1970, 5p.

Books

Boas, R.P.
Entire functions. Academic Press, New York, 1954.

Heins, M.
Selected topics in the classical theory of functions of a complex variable. Holt, Rinehart and Winston, New York 1962.

Helms, L.L.
Introduction to potential theory. Interscience 1969.

Landkof, N.S.
Foundations of modern potential theory. Nauka, Moscow 1966. Springer 1972.

Rudin, W.
Fourier analysis on groups. Interscience 1962.

Widder, D.V.
The Laplace transform, Princeton 1946.

Some recent papers

Baernstein theory.

Baernstein, A.

4. Integral means, univalent functions and circular symmetrization. Acta Math. 133, (1974), 133-169.

Fuchs, W.H.J.

2. A theorem on $\min_{|z|} \log|f(z)|/T(r,f)$. Symposium on Complex analysis, Canterbury (1973), LMS Lecture Note Series 12, 69-72.

Gariepy, R. and Lewis, J.

1. A maximum principle with applications to subharmonic functions in n-space. Ark. Mat. 12:2, (1974).

2. Space analogues of some theorems for subharmonic and meromorphic functions, to appear, Ark. Mat.

The last two papers extend the Baernstein theory to R^d, $d \geq 3$. Results of this type have also been obtained by Baernstein and Taylor.

Other papers

Hayman, W.K.

3. Functions subharmonic in the plane. Proc. of the Symposium in Mathematics at the Royal Institute of Technology in June 1973, Department of Mathematics, Royal Institute of technology, Stockholm, 57-68.

4. Research problems in function theory, Symposium on Complex Analysis, Canterbury 1973, LMS Lecture Note Series 12, 143-180.

Essén, M. and Jackson, H.

2. On the covering properties of a minimally thin set in a half-space, to appear.

The paper [3] of Hayman is a survey of the progress in this field during the last 20 years. In the paper of Essén and Jackson, the results of their first note [1] are extended from a Stolz domain to a half-space.

Index

An inequality for a class of harmonic functions in n-space

by

Christer Borell

1. Introduction

Let D be a bounded region in $\mathbb{R}^n$ for which the Dirichlet problem is solvable and let u be a solution of this problem with boundary data f. In this paper we shall give an upper estimate of u. More explicitely we perform symmetrizations of D and f and solve the corresponding Dirichlet problem. Let us call the solution u^0. We then try to find an upper estimate of u in terms of u^0. It shall be said at once that we make some restrictions on f.

We start with some definitions. Let L be an (n-k)-dimensional linear subspace of $\mathbb{R}^n$, $1 \leq k \leq n$. The k-dimensional Steiner symmetrization with respect to L then maps a compact subset M of $\mathbb{R}^n$ into a compact subset M^0 of $\mathbb{R}^n$ characterized by the following: If $a \in L$ and $(a + L^\perp) \cap M$ has positive k-dimensional Lebesgue measure, then $(a + L^\perp) \cap M$ and $(a + L^\perp) \cap M^0$ have the same k-dimensional Lebesgue measure and $(a + L^\perp) \cap M^0$ is a closed k-dimensional ball of centre a. However, if $(a + L^\perp) \cap M$ is a set of k-dimensional Lebesgue measure zero, then $(a + L^\perp) \cap M^0$ is empty or the singleton set $\{a\}$ according as $(a + L^\perp) \cap M$ is empty or non-empty.

Let $e_1,\dots,e_n$ be the standard basis in $\mathbb{R}^n$ and let the vector $x \in \mathbb{R}^n$ have the coordinates $x_1,\dots,x_n$ with respect to this basis.

In the following we assume that $L = [e_1,\dots,e_{n-k}]$, the linear subspace of $\mathbb{R}^n$ spanned by the first n - k basis vectors, and $1 \leq k \leq n-1$. Further, let $D \subseteq \mathbb{R}^n$ be a finite union of non-degenerated n-dimensional compact intervals, with sides parallel to the coordinate axes. We define $a_1(D),\dots,\gamma(D)$, by

$$a_i = \min\{x_i | x \in D\} \quad , \quad b_i = \max\{x_i | x \in D\}$$

$$, \quad i = 1,\dots,n-k$$

$$\alpha_i = D \cap \{x_i = a_i\} \quad , \quad \beta_i = D \cap \{x_i = b_i\}$$

$$\alpha = \bigcup_{1}^{n-k} \alpha_i \quad , \quad \beta = \bigcup_{1}^{n-k} \beta_i \quad , \quad \gamma = \alpha \cup \beta$$

Let D^o be the k-dimensional Steiner symmetrization of D with respect to L. For short, let us write $a_1(D^o) = a_1^o,\dots,\gamma(D^o) = \gamma^o$. Then, clearly, $a_i^o = a_i$ and $b_i^o = b_i$.

Let $f : \partial D \to [0, +\infty[$ be a function such that $f_{|\partial D \setminus \gamma} = 0$ and set for fixed i, $1 \leq i \leq n-k$,

$$M_{\alpha_i}(f) = \{x \in \mathbb{R}^n | a_i - f(x_1,\dots,x_{i-1},a_i,x_{i+1},\dots,x_n) \leq x_i \leq a_i ,$$

$$(x_1,\dots,x_{i-1},a_i,x_{i+1},\dots,x_n) \in \alpha_i\}$$

$$M_{\beta_i}(f) = \{x \in \mathbb{R}^n | b_i \leq x_i \leq b_i + f(x_1,\dots,x_{i-1},b_i,x_{i+1},\dots,x_n) ,$$

$$(x_1,\dots,x_{i-1},b_i,x_{i+1},\dots,x_n) \in \beta_i\}$$

In the following we shall assume that the sets $M_{\alpha_i}(f)$ and $M_{\beta_i}(f)$ are finite unions of n-dimensional compact intervals, with sides parallel to the coordinate axes.

It is easy to see that there exists a unique function $f^o : \partial D^o \to [0, +\infty[$ such that $f^o{}_{|\partial D^o \setminus \gamma^o} = 0$ and

$$(M_{\alpha_i}(f))^o = M_{\alpha_i^o}(f^o) \; , \; (M_{\beta_i}(f))^o = M_{\beta_i^o}(f^o) \; , \; i = 1,\dots,n-k \; .$$

In the following let us write $x = (x_1,\dots,x_{n-k} | x_{n-k+1},\dots,x_n) = (x',x'')$, $a' = (a_1,\dots,a_{n-k})$, and $b' = (b_1,\dots,b_{n-k})$. The notation $x' > a'$ then means that $x_i > a_i$, $i = 1,\dots,n-k$.

Now let $\omega(x,f,D)$ be the solution of the Dirichlet problem in D with boundary data f. It will be convenient to set $\omega(x,f,D) = 0$ if

$x \notin D$, $a' \leq x' \leq b'$.

The purpose of this paper is to prove the following.

Theorem 1.1: *If φ is any non-decreasing convex function on $\mathbb{R}$, then*

(1.1) $$\int \varphi(\omega(x,f,D))dx'' \leq \int \varphi(\omega(x,f^o,D^o))dx'' , \quad a' < x' < b' .$$

In particular,

(1.2) $$\max_{x''} \omega(x,f,D) \leq \omega((x',0''),f^o,D^o) , \quad a' < x' < b' .$$

Before the proof we will make a few remarks.

Above we have made several restrictions on D and f. The most important restrictions here are that D is bounded and that $f_{|\partial D \setminus \gamma} = 0$. Other than these restrictions Theorem 1.1 is obviously still true for much more general D and $f: \partial D \to [0, +\infty[$.

The characteristic function of a set $M \subseteq \mathbb{R}^n$ is denoted by χ_M. Note that if $f = \chi_\alpha$ or $f = \chi_\alpha + \chi_\beta$ in the Theorem 1.1, then $f^o = \chi_{\alpha^o}$ or $f^o = \chi_{\alpha^o} + \chi_{\beta^o}$, respectively. Theorem 1.1 is known in some special cases. In fact, Baernstein [1] proves the case $f = \chi_\alpha$, $n = 2$, $k = 1$, and Haliste [6, Theorem 8.2] proves the inequality (1.2) in the case $f = \chi_\alpha$, n arbitrary, $k = 1$. (Note, however, that Baernstein works with circular symmetrization with respect to the positive real axis.)

The inequality (1.1) is of the following general type;

$$\mu(\varphi) \leq \nu(\varphi) , \text{ all } \varphi', \varphi'' \geq 0 ,$$

where μ and ν are positive Borel measures on the real line both with compact support. Inequalities of this kind have been studied in great detail and the interested reader may consult [10], which also explains the underlying geometrical nature of the inequality.

Our proof of Theorem 1.1 is divided into two steps. In the first step we prove the special case $k = 1$ following closely to [1]. In fact, by defining a certain transform $u^*(x)$ of $u(x) = \omega(x,f,D)$, we can argue exactly as in [1]. The transform u^* has been introduced by Baernstein in the plane in a slightly different way, and with this definition u^* makes sense for any subharmonic function $u: \mathbb{R}^2 \to [-\infty,+\infty[$. The general case of Theorem 1.1 then follows from the special case $k = 1$ already proved by performing a certain smoothing process.

Finally, in this section, I wish to express my gratitude to Matts Essén and Peter Sjögren for many valuable discussions and suggestions.

2. The case $k = 1$.

If $f: \mathbb{R} \to \mathbb{R}_+$ is a measurable function, the function $\tilde{f}$ denotes the rearrangement of f in decreasing order on the interval $[0, +\infty[$.

We need the following

Lemma 2.1: For any $f, g, \in L^1_+(\mathbb{R})$ it holds that

$$\int_0^\infty |\tilde{f}(t) - \tilde{g}(t)|dt \leq \int_{-\infty}^\infty |f(t) - g(t)|dt .$$

This inequality is just a special case of the more general inequalities appearing in [3] and [8].

Lemma 2.2: Let E be a finite union of compact intervals. Then there is a positive number $\delta(E)$, which only depends on E, so that

$$(2.1) \quad \int_E f(t-a)dt + \int_E f(t+a)dt \leq \int_0^{|E|-2|a|} \tilde{f}(t)dt + \int_0^{|E|+2|a|} \tilde{f}(t)dt$$

for each $|a| < \delta(E)$, and each $f \in L^1_+(\mathbb{R})$.

Here $|E|$ denotes the one-dimensional Lebesgue measure of E.

Proof: Set $|(E-a) \cap (E+a)| = |E| - 2b$, $b \geq 0$. Then, clearly, the left-hand side of (2.1) is less or equal to

$$\int_0^{|E|-2b} \tilde{f}(t)dt + \int_0^{|E|+2b} \tilde{f}(t)dt ,$$

which is a non-increasing function of b. But since E is a finite union of compact intervals we have $b \geq |a|$ if $|a|$ is small enough. This proves Lemma 2.2.

Now suppose that $k = 1$ in Theorem 1.1 and set $u(x) = \omega(x,f,D)$ and $u_{x'}(x_n) = u(x',x_n)$. We define

$$u^*(x) = \int_0^{2x_n} \tilde{u}_{x'}(t)dt \ , \ a' \leq x' \leq b' \ , \ x_n \geq 0 \ .$$

Lemma 2.3: u^* is continuous in $a' \leq x' \leq b'$, $x_n \geq 0$ and subharmonic in $a' < x' < b'$, $x_n > 0$.

Proof: We first prove that u^* is continuous. Let $\bar{x} = (\bar{x}',\bar{x}_n)$ and $x = (x',x_n)$. Then

$$u^*(\bar{x}) - u^*(x) = u^*(\bar{x}) - u^*(\bar{x}',x_n) + \int_0^{2x_n} (\tilde{u}_{\bar{x}'}(t) - \tilde{u}_{x'}(t))dt \ .$$

Lemma 2.1 now gives

$$|u^*(\bar{x}) - u^*(x)| \leq \left| \int_{2x_n}^{2\bar{x}_n} \tilde{u}_{\bar{x}'}(t)dt \right| + \int |u_{\bar{x}'}(t) - u_{x'}(t)|dt$$

and the conclusion is obvious.

To prove that u^* is subharmonic let x be an arbitrary vector such that $a' < x' < b'$, $x_n > 0$. Since u is a real analytic function in the inerior of D we deduce that there exists a finite union of compact intervals E such that

$$u^*(x) = \int_E u_{x'}(t)dt \ , \ |E| = 2x_n \ .$$

But u is subharmonic in $a' < x' < b'$ and so we have the inequality

(2.2) $$u_{x'}(t) \leq \frac{1}{2} \left(\int (u_{x'+\rho y'}(t-\rho y_n) + u_{x'+\rho y'}(t+\rho y_n))d\sigma(y)\right) ,$$

valid for all t, and all $\rho < \min_{1 \le i \le n-1} (|a_i - x_i|, |b_i - x_i|)$. Here σ is the normalized surface-area measure on $\{|y| = 1\}$.

Now let us in addition assume that $\rho < \delta(E)$. By integrating (2,2) over E and making use of Lemma 2.2 we obtain

$$u^*(x) < \frac{1}{2} (\int (u^*(x'+\rho y', x_n - \rho y_n) + u^*(x'+\rho y', x_n+\rho y_n)) d\sigma(y)) ,$$

which proves that u^* is subharmonic.

Proof of Theorem 1.1, $k = 1$: Set $v(x) = \omega(x, f^o, D^o)$. To prove (1.1) it will be enough to show that $u^* \le v^*$. (Compare [7, Theorems 249 and 250].) To this end let us first define $D_+^o = D^o \cap \{x_n \ge 0\}$ and

$$v_1(x) = 2 \int_0^{x_n} v(x', t) dt , \quad x \in D_+^o .$$

The function v is clearly continuous and since $v(x', x_n) = v(x', -x_n)$ it follows that

$$\frac{\partial v}{\partial x_n} (x', 0) = 0, a' < x' < b' .$$

Therefore v_1 is harmonic in the interior of D_+^o. Furthermore, the functions v^* and v_1 are equal on ∂D_+^o. Since v^* is subharmonic we deduce that $v^* \le v_1$ in D_+^o. But the opposite inequality is obvious from the definitions so $v^* = v_1$ in D_+^o. In particular,

$$\frac{\partial v}{\partial x_n} \le 0 \text{ in the interior of } D_+^o .$$

We will now show that $u^* \le v^*$. Therefore let $\delta > 0$ be arbitrary, and set $w = u^* - v^* - \delta x_n$ in D_+^o. We know that w is continuous in D_+^o and subharmonic in the interior of D_+^o. It is also readily seen that $w \le 0$ on $(\partial D_+^o \cap \{x_n = 0\}) \cup (\gamma^o \cap \{x_n \ge 0\})$. Let $(p', p_n) \in \partial D_+^o$, $a' < p' < b'$, $p_n > 0$, and suppose that $w(p', p_n) \ge w(p', p_n - h)$ for all $h > 0$ small anough. We will show that this is impossible. Since u and v vanish on $\partial D \setminus \gamma$ it can be assumed that $(p', p_n + \delta) \notin \partial D_+^o$ for each $\delta > 0$.

Set $q_n = \max \{t | (p',t) \in D\}$. Then since $v^* = v_1$ in D_+^o , we have

$$\int_{-\infty}^{q_n} u(p',t)dt - 2 \int_0^{p_n} v(p',t)dt - \delta p_n \geq$$

$$\geq \int_{-\infty}^{q_n-2h} u(p',t)dt - 2 \int_0^{p_n-h} v(p',t)dt - \delta(p_n-h) ,$$

for all $h > 0$ small enough. This yields $u(p',q_n) \geq \delta/2$ which is impossible. We conclude that w cannot attain a maximum on the upper boundary of D_+^o and therefore $w \leq 0$. Since $\delta > 0$ is arbitrary, $u^* \leq v^*$ in D_+^o and so $u^* \leq v^*$ everywhere in $a' \leq x' \leq b'$. This proves (1.1) and (1.2) follows from (1.1) and the fact that $\frac{\partial v}{\partial x_n} \leq 0$ in the interior of D_+^o . This concludes the proof of Theorem 1.1 in the special case $k = 1$.

3. <u>The general case</u>.

In the following we assume that $n - 1 \geq k \geq 2$ and we set $g_1 = e_{n-k+1}, \ldots, g_k = e_n$. The linear subspace of $\mathbb{R}^n$ which is spanned by the vectors $g_1, \ldots, g_k$ will be denoted by $\mathbb{R}^k$. If A and B are non-empty compact subsets of $\mathbb{R}^k$ the Hausdorff distance $d(A,B)$ of A to B equals

$$\inf \{\delta > 0 \mid A + \delta\bigcirc \supseteq B\} + \inf \{\delta > 0 \mid B + \delta\bigcirc \supseteq A\} ,$$

where $\bigcirc$ is the closed unit ball in $\mathbb{R}^k$.

Let S denote the k-dimensional Steiner symmetrization in $\mathbb{R}^k$. A theorem due to Sarvas [9, Theorem 4.32] then says that there exist two (k-1)-dimensional Steiner symmetrizations in $\mathbb{R}^k$, S_1 and S_2 , respectively, such that

$$S(A) = \lim_{j\to\infty} (S_1 \circ S_2)^j(A)$$

for any compact non-empty subset A of $\mathbb{R}^k$. Using this we can quickly prove Theorem 1.1. However, the proof of Theorem 4.32 in [9] is rather lengthy and to do this paper independent of [9] we will prove a result (Lemma 3.2) which is similar to Sarvas' result. Our proof of Lemma 3.2 is based on other ideas than [9] and we hope the proof can be independent interest.

We need

Lemma 3.1: There exist k hyperplanes $H_1,\dots,H_k$ in $\mathbb{R}^k$ such that any convex body in $\mathbb{R}^k$ which is symmetric about each of these hyperplanes is a closed ball of centre 0 .

Here and in the following a convex body means a compact convex set with non-empty interior.

Lemma 3.1 is, of course, well-known, see e.g. [2, pp. 86] and [4, pp. 98].

Let us assume that the vector $y \in \mathbb{R}^k$ has the coordinates $y_1,\dots,y_k$ with respect to the basis $g_1,\dots,g_k$. Further, let $Q_1,\dots,Q_k$ be isometric operators such that

$$Q_i(H_i) = \{y_k = 0\} \ , \ i = 1,\dots,k \ ,$$

where $H_1,\dots,H_k$ are as in Lemma 3.1. We let S_i denote the one-dimensional Steiner symmetrization with respect to the hyperplane $y_i = 0$ and set $S = S_1 \circ \dots \circ S_k$.

Lemma 3.2: Let A be a compact subset of $\mathbb{R}^k$ with non-empty interior and define

$$A_1 = S(A) \ , \ A_{j+1} = Q_j^{-1}\ (S(Q_j(A_j))) \ , \ j \geq 1 \ ,$$

where $Q_{k+1} = Q_1$, $Q_{k+2} = Q_2,\dots$.

Then $A_j \to \rho O$ as $j \to +\infty$, where $\rho > 0$ fulfils $m_k(\rho O) = m_k(A)$.

Here m_k is the k-dimensional Lebesgue measure.

If A is a convex body, Lemma 3.2 is a well-known result due to Blaschke. (see e.g. [2, pp. 86] and [4, pp. 98].) Our proof of Lemma 3.2 uses convexity methods and is a modification of the classical proof in the case when A is a convex body. The proof leans heavily on the fact that $d(\bar{\bar{A}},\bar{\bar{B}}) \leq d(A,B)$, where $\bar{\bar{A}}$ denotes the convex hull of A . This inequality follows directly from the identity $\overline{\overline{A+B}} = \bar{\bar{A}} + \bar{\bar{B}}$, which is valid for any subsets A and B of $\mathbb{R}^k$. Before the proof of Lemma 3.2 let us formulate

Definition 3.1. A subset A of $\mathbb{R}^k$ is said to be a k-convex symmetric body if

1) A is compact with non-empty interior.
2) A is symmetric about each hyperplane $y_i = 0$, $i = 1,\ldots,k$.
3) The intersection of A and any straight line parallel to a coordinate axis is an interval.

We need a simple

Lemma 3.3:

a) If A is any compact subset of $\mathbb{R}^k$ with non-empty interior, then S(A) is a k-convex symmetric body.
b) If a sequence of k-convex symmetric bodies converges, then the limit set fulfils the conditions 2) and 3) of Definition 3.1.
c) If a sequence $\{A_j\}_1^\infty$ of k-convex symmetric bodies converges to a closed ball B of centre 0 , then $m_k(A_j) \to m_k(B)$ as $j \to +\infty$.

Proof: Everything here is trivial but let us prove c).

Note first that $\bar{\bar{A}}_j \to \bar{\bar{B}} = B$ as $j \to +\infty$. Hence $\overline{\lim_{j\to\infty}}\, m_k(A_j) \leq m_k(B)$.

It can be assumed that $m_k(B) > 0$ and let y belong to the interior of B. We claim that $y \in A_j$ if j is large enough. It can clearly be assumed that $y_1,\ldots,y_k \geq 0$. Now choose $\delta > 0$ such that

$$[(\mathbb{R}^k \setminus \prod_1^k [y_k,+\infty[) + \delta O] \not\supseteq B .$$

If $y \notin A_j$, then Definition 3.1 implies that

$$(A_j + \delta O) \not\supseteq B .$$

Therefore, we conclude that $y \in A_j$ if j is large enough. Fatou's lemma now gives

$$m_k(B) \leq \underline{\lim}_{j\to\infty} m_k(A_j) ,$$

which proves c).

Proof of Lemma 3.2: Let $(j_p)_1^\infty$ be a strictly increasing sequence of natural numbers such that $A_{j_p} \to B$ as $p \to +\infty$. By the Blaschke selection theorem [5, p. 154] it suffices to prove that $B = \rho O$.

First note that there exist a natural number r , $1 \leq r \leq k$, and a strictly increasing sequence $(p_\nu)_1^\infty$ of natural numbers such that $Q_{j_{p_\nu}} = Q_r$, and $j_{p_{\nu+1}} - j_{p_\nu} \geq k$ for all ν .

The surface-area of a convex body M is denoted by $\sigma(M)$. The Cauchy surface-area formula then implies that $\sigma(M_1) \leq \sigma(M_2)$ for all convex bodies M_1 and M_2 such that $M_1 \subseteq M_2$. Furthermore, if M is any compact subset of $\mathbb{R}^k$, we have $\overline{\overline{S(M)}} \subseteq S(\overline{\overline{M}})$. Using these two properties, we get

$$(3.1) \qquad \sigma(\bar{\bar{A}}_{j_{p_{\nu+1}}}) \leq \sigma(Q^{-1}_{j_{p_{\nu+1}}-1}(S(Q_{j_{p_{\nu+1}}-1}(\bar{\bar{A}}_{j_{p_{\nu+1}}-1})))) \leq$$

$$\leq \sigma(\bar{\bar{A}}_{j_{p_{\nu+1}}-1}) \leq \dots \leq \sigma(Q_r^{-1}(S(Q_r(\bar{\bar{A}}_{j_{p_\nu}})))) .$$

By letting $\nu \to +\infty$ it follows that

$$(3.2) \qquad \sigma(\bar{\bar{B}}) \leq \sigma(Q_r^{-1}(S(Q_r(\bar{\bar{B}})))) = \sigma(S(Q_r(\bar{\bar{B}}))) .$$

Hence $\sigma(Q_r(\bar{\bar{B}})) \leq \sigma(S_k(Q_r(\bar{\bar{B}})))$, and we deduce that $Q_r(\bar{\bar{B}}) = S_k(Q_r(\bar{\bar{B}}))$. Using (3.2), we conclude that $\sigma(Q_r(\bar{\bar{B}})) \leq \sigma(S_1 \circ \dots \circ S_{k-1}(Q_r(\bar{\bar{B}}))) \leq$ $\leq \sigma(S_{k-1}(Q_r(\bar{\bar{B}})))$. By repetition, we get $Q_r(\bar{\bar{B}}) = S(Q_r(\bar{\bar{B}}))$.

Now observe that $\bar{\bar{A}}_{j_{p_{\nu+1}}} \subseteq Q_r^{-1}(S(Q_r(\bar{\bar{A}}_{j_{p_\nu}})))$. The inequality (3.1) thus gives

$$\sigma(\bar{\bar{A}}_{j_{p_{\nu+1}}}) \leq \sigma(Q_{r+1}^{-1}(S(Q_{r+1}(\bar{\bar{A}}_{j_{p_\nu}+1})))) \leq$$

$$\leq \sigma(Q_{r+1}^{-1}(S(Q_{r+1}(Q_r^{-1}(S(Q_r(\bar{\bar{A}}_{j_{p_\nu}}))))))) .$$

By letting $\nu \to +\infty$ we get

$$\sigma(\bar{\bar{B}}) \leq \sigma(Q_{r+1}^{-1}(S(Q_{r+1}(\bar{\bar{B}})))) ,$$

since $\bar{\bar{B}} = Q_r^{-1}(S(Q_r(\bar{\bar{B}})))$. By the same argument as above it follows that $Q_{r+1}(\bar{\bar{B}}) = S_k(Q_{r+1}(\bar{\bar{B}}))$ and $Q_{r+1}(\bar{\bar{B}}) = S(Q_{r+1}(\bar{\bar{B}}))$. By repetition, we have that the sets $Q_i(\bar{\bar{B}})$, $i = 1,\dots,k$, are symmetric about the hyperplane $y_k = 0$, and lemma 3.1 proves that $\bar{\bar{B}}$ is a closed ball of centre 0 .

In the next step it will be proved that $\bar{\bar{B}} = B$. Suppose to the contrary that $\bar{\bar{B}} \setminus B \neq \emptyset$. Since B is closed there exists a $\delta > 0$ such that $\bar{\bar{B}} \setminus (B + 2\delta\bigcirc) \neq \emptyset$. Choose j_p so that $A_{j_p} + \delta\bigcirc \supseteq B$ and $B + \delta\bigcirc \supseteq A_{j_p}$. Then $B + 2\delta\bigcirc \supseteq A_{j_p} + \delta\bigcirc$ and $\bar{\bar{B}} \setminus B \supset \bar{\bar{B}} \setminus (A_{j_p} + \delta\bigcirc) \neq \emptyset$. Since $Q_{j_p-1}(A_{j_p} + \delta\bigcirc) = Q_{j_p-1}(A_{j_p}) + \delta\bigcirc$ is a k-convex symmetric body we have a contradiction. Hence $\bar{\bar{B}} = B$. This also shows that $d(Q_{j_p-1}(A_{j_p}), B) = d(A_{j_p}, B) \to 0$ as $p \to \infty$. Lemma 3.3c) therefore gives $B = \rho\bigcirc$, which was to be proved.

<u>Proof of Theorem 1.1</u>: We are now working in $\mathbb{R}^n$. Let $\bar{S}_i$ denote the one-dimensional Steiner symmetrization with respect to the hyperplane $x_{i+n-k} = 0$ for $i = 1,\dots,k$, and set $\bar{S} = \bar{S}_1 \circ \dots \circ \bar{S}_k$. Then there exists a unique function

$$\bar{S}_k(f) : \partial(\bar{S}_k(D)) \to [0, +\infty[,$$

vanishing on $\partial\overline{S}_k(D)\setminus\gamma(\overline{S}_k(D))$ and such that

$$\overline{S}_k(M_{\alpha_i}(f)) = M_{\overline{S}_k(\alpha_i)}(\overline{S}_k(f)) \text{ , } \overline{S}_k(M_{\beta_i}(f)) = M_{\overline{S}_k(\beta_i)}(\overline{S}_k(f)) \text{ ,}$$

$i = 1,\dots,n-k$.

The functions $\overline{S}_{k-1} \circ \overline{S}_k(f) = \overline{S}_{k-1}(\overline{S}_k(f)),\dots,\overline{S}(f)$ are defined in the same way.

In the following let φ be an arbitrary non-decreasing convex function on $\mathbb{R}$. The special case $k = 1$ of Theorem 1.1 then yields

$$\int \varphi(\omega(x,f,D))\, dx_n \leq \int \varphi(\omega(x,\overline{S}_k(f),\overline{S}_k(D)))\, dx_n \text{ ,}$$

$$\int\varphi(\omega(x,\overline{S}_k(f),\overline{S}_k(D)))dx_{n-1} \leq \int\varphi(\omega(x,\overline{S}_{k-1} \circ \overline{S}_k(f),\overline{S}_{k-1} \circ \overline{S}_k(D)))dx_{n-1} \text{ ,}$$

$$\vdots$$

$$\int\varphi(\omega(x,\overline{S}_2\circ\dots\circ\overline{S}_k(f),\overline{S}_2\circ\dots\circ \overline{S}_k(D)))dx_{n-k+1} \leq \int\varphi(\omega(x,\overline{S}(f),\overline{S}(D)))dx_{n-k+1} \text{ .}$$

Hence

$$\int \varphi(\omega(x,f,D))dx'' \leq \int \varphi(\omega(x,\overline{S}(f),\overline{S}(D))dx'' \text{ .} \tag{3.3}$$

Now let the isometric operators $Q_1,\dots,Q_k$ be as above. These operators have unique isometric extensions to $\mathbb{R}^n$, denoted by $\overline{Q}_1,\dots,\overline{Q}_k$, respectively, so that $\overline{Q}_i(e_j) = e_j$, $i=1,\dots,k$, $j=1,\dots,n-k$.

If $h\colon M \to [0,+\infty[$ $(M \subseteq \mathbb{R}^n)$ and $T\colon \mathbb{R}^n \to \mathbb{R}^n$ is a 1-1 linear operator let Th be the mapping from $T(M)$ into $[0,+\infty[$ defined by $Th(x) = h(T^{-1}x)$, $x \in T(M)$.

We define

$$D_1 = \overline{S}(D) \text{ , } D_{j+1} = \overline{Q}_j^{-1}(\overline{S}(\overline{Q}_j(D_j))) \text{ , } j \geq 1 \text{ ,}$$

$$f_1 = \overline{S}(f) \text{ , } f_{j+1} = \overline{Q}_j^{-1}(\overline{S}(\overline{Q}_j(f_j))) \text{ , } j \geq 1 \text{ .}$$

Note that

$$\omega(\overline{Q}_1^{-1}x,f_1,D_1) = \omega(x,\overline{Q}_1(f_1),\overline{Q}_1(D_1)) .$$

Therefore (3.3) implies that

$$\int\varphi(\omega(\overline{Q}_1^{-1}x,f_1,D_1))dx'' \leq \int\varphi(\omega(x,\overline{S}(\overline{Q}_1(f_1)),\overline{S}(\overline{Q}_1(D_1)))dx'' .$$

But

$$\int\varphi(\omega(\overline{Q}_1^{-1}x,f_1,D_1))dx'' = \int\varphi(\omega(x,f_1,D_1))dx''$$

and a comparison with (3.3) gives

$$\int\varphi(\omega(x,f,D))dx'' \leq \int\varphi(\omega(x,f_2,D_2))dx'' .$$

By repetition we obtain

$$\int\varphi(\omega(x,f,D))dx'' \leq \int\varphi(\omega(x,f_j,D_j))dx'' , \quad j \geq 1 . \tag{3.4}$$

Lemma 3.2 shows that $L^\perp \cap (D_j + a) \to L^\perp \cap (D^o + a)$ as $j \to +\infty$ for each $a \in L_\perp$. For fixed j the set-valued functions $a \to L^\perp \cap (D_j + a)$ and $a \to L^\perp \cap (D^o + a)$ only attain a finite number of sets and they have the same points of discontinuity. Hence $D_j \to D^o$ as $j \to \infty$.

Now let

$$M(f) = \bigcup_1^{n-k} (M_{\alpha_i}(f) \cup M_{\beta_i}(f)) ,$$

where $M_{\alpha_i}(f)$ and $M_{\beta_i}(f)$ are as in the introduction. It is then readily readily seen that

$$M(f_1) = \overline{S}(M(f)), M(f_{j+1}) = \overline{Q}_j^{-1}(\overline{S}(\overline{Q}_j(M(f_j)))) , \quad j \geq 1 .$$

Using Lemma 3.2 again we get that $M(f_j) \to M(f^o)$ as $j \to +\infty$. By letting $j \to +\infty$ in (3.4) we obtain (1.1). The inequality (1.2) follows in a similar way. This proves Theorem 1.1.

References

1. Baernstein II, A.: Integral means, univalent functions and circular symmetrization, Acta Math. 133 (1974), 139-169.

2. Blaschke, W.: "Kreis und Kugel," Walter De Gruyter, Berlin, 1956.

3. Borell, C.: A note on an inequality for rearrangements, Pacific J. Math., Vol. 47, No. 1 (1973), 39-41.

4. Eggleston, H.G.: "Convexity", Cambridge Tracts, Cambridge, 1958.

5. Hadwiger, H.: "Vorlesungen über Inhalt, Oberfläche und Isoperimetrie" Springer-Verlag, 1957

6. Haliste, K.: Estimates of harmonic measures, Ark. Mat. 6 (1965-67) 1-31.

7. Hardy, G.H., Littlewood, J.E., and Pólya, G.: "Inequalities", 2nd ed., Cambridge University Press, London, 1964.

8. Lorentz, G.G.: An inequality for rearrangements, Amer. Math. Monthly, 60 (1953), 176-179.

9. Sarvas, J.: Symmetrization of condensers in n-space, Ann. Acad. Sci. Fennicae 522 (1973).

10. Strassen, V.: The existence of probability measures with given marginals, Ann. Math. Statist., 36 (1965), 423-439.